the clever guts diet

the clever guts diet

SB

The information contained in this book is provided for general purposes only. It is not intended as and should not be relied upon as medical advice. The publisher and authors are not responsible for any specific health needs that may require medical supervision. If you have underlying health problems, or have any doubts about the advice contained in this book, you should contact a qualified medical, dietary or other appropriate professional.

Published in 2017 by
Short Books, Unit 316, ScreenWorks, 22 Highbury Grove,
London, N5 2ER

10 9 8 7 6 5 4 3 2 1

A CIP catalogue record for this book
is available from the British Library.

ISBN: 978-1-78072-304-4

Cover design by Andrew Smith
Printed at CPI Group (UK) Ltd, Croydon, CR0 4YY

'All disease begins in the gut.'

Hippocrates of Kos,
the father of modern medicine

CONTENTS

Introduction

Although this book has 'diet' in its title, it is not really about losing weight. That may well happen, if you eat the foods and do the things I recommend, but that's not its primary purpose. *The Clever Guts Diet* is a 'diet' in the same way you might talk about being on a vegetarian diet or a Mediterranean diet. It's not about calories or restriction; it's about the sort of food and lifestyle changes you should make if you have gut problems, or simply want to keep yours in good condition.

The gut is not a glamorous organ. When I was at medical school many of my fellow students wanted to study the brain, by training as neurosurgeons, or become cardiologists, experts in the heart. I never heard anyone say they wanted to dedicate their life to the gut. And yet it is extraordinary – a part of the body, hitherto relatively unexplored, with which I have recently become rather obsessed. Thanks to a huge amount of new research, probing the world within our guts is changing our understanding of the way our bodies work.

As well as extracting energy from our food, the gut accounts for most of our immune system and produces more than two dozen hormones that influence everything from our appetite to our mood.

I also love the fact that, buried in our intestines, deep inside its tissue, is a very thin layer of brain. It's called the enteric system and it is made up of the same cells, neurons, which are found in the brain. There are over 100 million neurons in the gut, as many as you would find in the brain of a cat. Except, instead of being in one big lump, like the brain on top of your neck, the neurons in your gut are spread out in a thin mesh that extends all the way from your throat to your rectum. This 'second brain' doesn't do much geometry or worry about tax returns, but it does orchestrate digestion and moderate gut pain.

When we talk about having 'gut feelings' or 'gut instincts' we are reflecting the reality of how closely our guts and are brains are entwined. In this book I am going to be talking a lot about the 'gut–brain axis' and the new science that surrounds it.

Your gut is a wonderful piece of engineering and I hope, after you have read this, you will share my enthusiasm for it. But in many ways the star of the digestive show is not actually part of the human body at all – it is the one to two kilos of microbes that live in your gut and make up the microbiome.

Until recently the world of the microbiome was a dark, dank and private one. Down there live creatures that have never seen the light of day, more than 50 trillion of them, at least 1000 different species, a richer diversity of life than you would find in a rainforest.

As is often the case with new scientific discoveries, a lot of genuine research has been misinterpreted and findings have been exaggerated. And, just as these microbes were previously ignored, now they are in danger of being seriously overhyped. Recent research shows that we are not '90 per cent bacteria' and '10 per cent human', as many books and articles have claimed, but more like 50:50.[1]

In fact, according to one of the researchers who helped explode this myth, the proportions are so similar that 'each defecation may flip the ratio to favour human cells over bacteria'.

More importantly, while there are foods that will help your microbiome thrive (and that's why this book contains recipes), few of the products that are sold on this basis have credible science behind them. When it comes to prebiotics, probiotics and supplements, I will show you what works, and what doesn't.

Our widespread ignorance about the microbiome arises from the fact that, until quite recently, its inhabitants, microbes, were impossible to study. We knew they helped protect the gut from dangerous invaders; that they synthesised a few vitamins; and that they gobbled up fibre that our bodies can't digest.

Now we know they do far more than that:

1. They help regulate our body weight. As we'll see in later chapters the microbes in your gut can decide how much energy your body extracts from the food you eat; they control hunger signals; they help decide which foods you crave; and they determine how

much your blood sugar spikes in response to a meal. Can your microbiome make you fat? It certainly can. Can you change your microbiome so it works with you rather than against you? You certainly can, and I will show you how.

2. The microbiome not only protects our guts from invaders, it teaches and regulates our entire immune system. Over the last half-century we have seen a massive rise in allergic diseases, such as asthma and eczema, caused by an overactive immune system. We have also seen a huge surge in autoimmune diseases, ranging from inflammatory bowel disease to type 1 diabetes, which again are primarily caused by an immune system that has got out of control. I will show you how changing the mix of bacteria in your gut can reduce the impact of these diseases.

3. The microbiome takes the bits of food our body can't digest and converts them into a wide range of hormones and chemicals. These, it seems, can control our mood, as well as our appetite and general health. Changing your biome may reduce anxiety and lessen depression.

The tragedy is that, in our ignorance, we have been laying waste to our microbiome and its population of microbes, or 'Old Friends'. They've been given that name because they have evolved with us over millions of years, and also because so many of them are essential to our health. Just as we have ravaged the rainforests and consigned numerous animal species to oblivion, so we have decimated the populations that live inside us. Fortunately we can help these Old Friends bounce back. I will show you how.

I will also be looking into the latest treatments for a number of gut problems, ranging from gluten intolerance to irritable bowel syndrome. These are diseases that many people struggle with, in part because doctors are often bad at both diagnosing and treating them. They are frequently dismissed as 'psychosomatic' – that is, the product of anxiety or depression.

The same used to be said of stomach ulcers. Also known as gastric ulcers, these are open sores that develop on the lining of the stomach and small intestine.

Back in 1994, when I made a television programme about ulcers (which I unimaginatively called *Ulcer Wars*), they were common and considered incurable. It was widely believed that they were caused by stress, which made your stomach produce too much acid, and that was what did the damage. The standard medical advice was to eat bland food, change your stressful lifestyle and take a drug to reduce acid production. If that didn't work, and it often didn't, you might find yourself in the hands of surgeons having parts of your stomach and small bowel removed.

But in Perth, Western Australia, there were a couple of doctors who did not think that stress was the real cause of ulcers. They argued that most ulcers are the result of infection by a previously unknown bacterium that they had identified and named *Helicobacter Pylori.*

To make his point, in 1984 one of the scientists, Dr Barry Marshall, brewed up a flask of *Helicobacter* and swallowed it. A few days later, as he smilingly told me, he started vomiting. He had himself endoscoped; a small tube was passed down his throat and into his stomach. Samples of his now inflamed stomach lining were removed. These showed that his stomach had been colonised by *Helicobacter.*

Barry's wife, Robin, worried that he would become seriously ill, insisted that he stop the experiment. So Barry took a handful of antibiotics, which he had previously shown could kill *Helicobacter*, and soon his stomach was back to normal.

Ten years later, and despite extensive research showing that a short course of antibiotics could cure stomach ulcers, most of the experts I interviewed for my film dismissed Barry's work out of hand. One told me he refused to believe a major breakthrough could have come out of an 'academic backwater like Perth'. A gut expert who reviewed my film in the *British Medical Journal* described it as 'one sided and tendentious'.

Normally you make a documentary, it goes out, and that is that. Not with *Ulcer Wars.* I received tens of thousands of letters (these were the days before the internet or email) from people in terrible pain, who had not responded to standard treatments. I ended up sending out thousands of fact sheets, describing the science and Barry's antibiotics protocol.

I still have some of the letters I got back, including one from a guy called Brian whose ulcer was not responding to standard treatment and who had been told to give up the high-powered job he loved and have most of his stomach removed. He took my fact sheet to his doctor, and begged for a course of antibiotics. His doctor reluctantly agreed and within a couple of weeks he was completely cured. He wrote to me regularly afterwards to say he is still going strong.

The tide slowly turned and I was delighted when Barry Marshall and Robin Warren won the Nobel Prize for Medicine for their work in 2004. Looking for and treating *Helicobacter* infection when people have gastric ulcers is now completely standard practice.

The point I'm making is not that antibiotics are the solution to everything. They aren't and their overuse has created other serious gut problems. Nor am I suggesting that stress doesn't matter. It does, and I will show you proven ways to de-stress.

The point is that many diseases have been dismissed as psychosomatic simply because doctors haven't had the right tools to investigate them properly. In the 1930s asthma was treated by psychotherapy because of the mistaken belief that it was 'all in the mind'. Autism and schizophrenia were once blamed on poor parenting.

One reason I've written this book is that I'm convinced that many common gut conditions are better treated by a change in diet than by drugs or antidepressants.

The initial chapters provide an overview of the gut,

which I have written in the form of a journey through my own intestines. These chapters include not only what the gut does, but also what happens when it goes wrong.

Chapter 3 introduces us to the wonderful world of the microbiome and some of the more influential tribes that you will find down there.

Subsequent chapters look at the unexpected ways in which our microbiome influences us, before moving on to scientifically tested ways to keep it in good shape. And finally there is a section of recipes from nutritional therapist Tanya Borowski and GP Clare Bailey.

I've learnt a great deal that has surprised me, and got such a lot of practical benefit from researching this book. I now eat a much wider range of foods, including fermented foods I'd never tried before. I've thoroughly enjoyed the journey. I hope you do too.

PART I

1

Down the Hatch

As I stood there in the Science Museum in London I began to wonder if this really was such a great idea. In a moment of exuberance I had agreed to take part in a live event at the world-famous museum. I'd said I would swallow a small pill-shaped camera, linked via sensors on my body to a giant screen so that an audience of hundreds of people would be able to explore, along with me, the intimate, alien world of my guts.

During my medical training and more recently as a television presenter, I have taken part in some pretty bizarre and painful experiments, but nothing quite like this. In preparation I had fasted for 36 hours and taken some powerful laxatives. This was to ensure that the camera would have as clear a view as possible when it got to the murkier parts of my digestive system.

I realised that once I had swallowed the camera I would have no further control of it – how long it would take to work its way through my guts was anybody's guess. We know that food can take up to three days to travel from lips to exit, sometimes ever longer. I expected that, because

I had drunk four litres of laxative the night before, my pill cam would travel rather more quickly than normal. But my audience were warned that if they were waiting for the camera to pop out the other end they could be in for a very long wait.

The camera I had to swallow was smooth and oblong, a bit more than a centimetre long, the size of a large vitamin pill. Considering the fact that it houses the equivalent of a film crew, complete with lights and camera, it is impressively small. But still a bit of a gulp. It's a bit of kit that is normally used by gastroenterologists to go to the parts of their patients that endoscopes and colonoscopes can't reach.

When everything was set up, I swallowed the pill cam, with the help of a glass of water. Down it went, past my tonsils and into my oesophagus, beaming live pictures as it travelled. When you swallow food, the oesophagus detects it as it touches its wall and the muscles begin to contract, pushing the food further down. The contraction is so powerful that in theory you could eat while standing on your head.

That was what was supposed to happen. In fact, the camera got stuck at the bend where the oesophagus meets the stomach.

I had a moment of panic when I wondered whether I would end up in surgery having my gullet operated on, but fortunately after jumping up and down on the spot for a few moments the camera navigated the narrowing and dropped into my stomach.

Seeing your stomach from the inside, via a fish-eye lens, is an unusual experience. Down there is a cavernous, exotic landscape – pulsing and throbbing with movement. It was pink and pulsatile, a place of slobber and rawhide. When it is empty, the mucosa that lines the stomach is thrown up in folds like a boggy marsh. It reminded me of the surface of Mars. Except slimier. And much more active.

The walls of the stomach are constantly on the move, contracting and folding in on themselves. If the camera had been digestible it would have been pounded and mashed into fragments, then dunked in gastric juices as acidic as a car battery. This acid bath is there to destroy any harmful bacteria or parasites that you swallow along with your food.

In some people, the highly acidic contents of the stomach can leak back up into the oesophagus. Known as acid reflux, this can be very painful as it tends to burn the lining. It is commonly treated with acid-suppressing drugs. (If symptoms are mild, ginger tea can help – infuse a couple of slices of fresh ginger in hot water for about 30 minutes. Drink it before a meal.)

Interestingly, it wasn't a camera that gave us the first real insights into our digestive workings. It was, instead, a terrible accident.

The story begins in June 1822, when a young Canadian boatman, Alexis St Martin, working on the shores of Lake Michigan, was accidentally shot in the chest. The blast ripped through his ribs, his lungs and the front wall of his stomach. Parts of his undigested breakfast started pouring out of his body, along with bits and pieces of his torn stomach. The first person on the scene was a young army doctor called William Beaumont. He dressed the wound, but things didn't bode well and he really didn't expect Alexis to live. Still, survive he did – and in doing so he laid the foundations of our modern understanding of the guts.

Alexis's original wound was huge, about the size of a man's palm, high up on the left side of his chest. We tend to think of the stomach as lying somewhere in the middle, down by the belly button, as that is where a lot of the gurgling noises and stomach pains seem to come from, but it's actually much higher up, right below the diaphragm. The giant hole in his chest shrank, but never fully closed, leaving an opening leading directly from the outside into his stomach. An opening, or hole, like this is known as a fistula.

This was obviously unfortunate for Alexis St Martin, but a great opportunity for William Beaumont to study the living digestive system in a way that no other surgeon or physician had ever been able to do before.

To keep Alexis close, Beaumont employed him as his handyman and embarked on a series of investigations that went on for nearly 10 years.

He would wrap up food in muslin bags – bits of vegetable or meat – and then pop them directly into Alexis's stomach via the fistula. He would leave the bag for a while, then pull it out to see what had happened. A bit like brewing a cup of tea.

As well as putting things into the hole in Alexis's stomach, Beaumont sucked juices out. And it was these experiments that proved to be truly revolutionary, because they turned popular beliefs about digestion on their head.

At that time, the early years of the 19th century, it was assumed that digestion was purely mechanical, the muscles in the stomach mashing up the food. Beaumont showed that this was not true. Digestion is also a chemical process. He discovered, for example, that the juice he sucked out from Alexis's stomach was full of hydrochloric acid, which is highly corrosive. He also discovered that these juices were full of digestive enzymes that could break down food if they were mixed together in a bowl.

What happens when you eat and drink

I was keen to see some of this activity going on in my own stomach, so with the camera still in place I ate a meal of steak, chips and vegetables, washed down by a glass of apple juice.

I was soon watching, along with my enthralled and slightly disgusted audience, the food and drink that I'd just swallowed arrive in my stomach. The liquid swiftly ran down the walls to the bottom of the stomach, where it joins the small intestine. This is known as the gastroduodenal junction. Down there is a muscular valve, the pyloric sphincter, which acts like a bouncer at a night club, deciding who or what is allowed to pass. When the sphincter is contracted, closed, it holds the food you've eaten in your stomach, allowing the digestive juices and mechanical pounding to do their work. Once it decides that something has been sufficiently pulverised, it opens up and allows it to pass into your duodenum. The whole process has to be carefully orchestrated, which is why you have so many brain cells down there.

Anyway, after a brief pause, the pyloric sphincter decided the apple juice I'd drunk was OK, so it was allowed to pass into my small intestine, like water emptying out of the bath.

There is a useful lesson here for anyone who wants to lose weight: don't drink your calories.

Unlike those in food, the calories we consume in the form of drink don't fill us up, and some of the worse offenders are the treats which people fondly imagine are healthy. In particular we've been sold the idea that fruit juice and smoothies are good for us because they come from a natural source and contain vitamins. The truth is, unless they are freshly made, they are not a great source of nutrients. What they are good at – the main reason they are so popular – is delivering a very rapid sugar hit to your brain.

You drink the juice, it goes down the hatch, swiftly through your stomach and into your small intestine where the sugar is extracted and absorbed into your bloodstream and transported to your brain. Your brain releases the 'feel good' hormone, dopamine, and you get that familiar sugar kick. Yeah! Give me more of that stuff.

The surplus energy from all that sugar has to go somewhere, and unless you burn it off quickly by doing lots of exercise, it will be stored as fat, either in your liver or around your gut. There are around 120 calories in a small cup or carton of apple or orange juice, the equivalent of five teaspoons of sugar. It may be 'natural' but it is still sugar and your body will treat it like Coca Cola. Shop-bought smoothies have similar levels of sugar to fruit juice, some even higher. Drop for drop, grape juice is the worst of the juices, with a single cup containing as much sugar as four doughnuts.

Can you run it off through exercise? Yes, but many people underestimate how much extra exercise they have to do to burn off the calories contained in a small treat.

You may think, 'well, in that case I'll have one of those drinks with zero-calorie sweeteners that I've seen advertised everywhere.' Sadly, the evidence is mounting that drinking these can lead to inflammation in the gut and an increased risk of obesity. (For more on this, see page 152).

It is far better to train your taste buds to go for something less sweet. When I'm thirsty I stick to tea, coffee, herbal tea and water. I love chilled fizzy water with flavourings like lemon, lime juice or just a slice of cucumber.

Although I am not a fan of fruit juice or smoothies, I do enjoy eating fruit, particularly apples and pears, which are less sweet. One of the big differences between eating an apple and drinking it in the form of apple juice is that an apple has far more nutrients and a lot more fibre, so it will hang around inside your stomach for a lot longer and you won't get such a big sugar rush. The undigested bits of the apple will also pass through your small intestine and down into your large intestine, where they will help feed your 'good' bacteria. More on that later.

The simple message is that while eating an apple will fill you up, drinking apple juice will leave you hungrier and fatter.

Alcohol

While we are on the subject of drink, what about alcohol? What happens when you knock back a glass of wine or beer? Like any drink it swiftly passes into your stomach. If your stomach is empty then the alcohol will irritate the lining, making the blood vessels swell, increasing both the speed with which the alcohol is absorbed into the blood and the amount of it that goes in.

If you have had something to eat before you drink, particularly something fatty, it will line your stomach, acting as a physical barrier to the booze. So instead of passing straight into the blood vessels that line your stomach, the alcohol will pass out through the pyloric sphincter and into your small intestine.

The small intestine is where most alcohol is absorbed. If it goes into your blood by this route, it takes longer to get into your bloodstream, and so to your brain. If you drink on an empty stomach, your brain typically gets a peak alcohol hit somewhere between 30 and 60 minutes later. If you eat something fatty before you start drinking, that peak will be delayed by up to an hour.

Once alcohol hits your brain, it depresses your inhibitory centres. Contrary to what it may feel like, alcohol is not a stimulant. When I drink too much I normally have a short period of uninhibited fun, follow by a slump and a strong desire to go to sleep. Drinking more than a glass or two of wine doesn't make me good company for long. To slow my drinking, I aim to alternate a glass of water with every glass of wine.

With alcohol in your system, your liver now gets to work, busily breaking it down. First it turns it into a mildly toxic substance called acetaldehyde. This then gets converted into vinegar (acetic acid) and eventually into carbon dioxide and water.

It is the build-up of acetaldehyde (which is more toxic than alcohol) that causes flushing and makes hangovers such a painful experience.

Your liver can break down roughly one unit of alcohol an hour. Drink faster than that and you will get smashed. Keep on drinking heavily and you will eventually destroy your liver.

My wife is not a big drinker, which is just as well. If a woman decides to try and match a man drink for drink, she will normally get drunker, more quickly. That is partly because women tend to be smaller, but also because they have less muscle and more body fat. Some alcohol gets absorbed by your muscles, and thus removed from your system, but fat and alcohol simply don't mix.

Hormones also play a part in metabolising alcohol, and

a woman who goes drinking just before she has her period will find less booze will have the same effect.

Does mixing drinks make you drunker? Broadly, no. How drunk you get is mainly down to how much alcohol you drink, though fizzy drinks are an exception. If you decide to start your evening with a glass of champagne (or sparkling wine), the carbon dioxide in the drink will open the pyloric sphincter valve linking your stomach to your small intestine, allowing the alcohol to pass into your system faster, and getting you drunk more quickly.

Although I have given up fruit juice, I still drink alcohol. That's not only because I like it, but because I believe there are health benefits to be had from moderate drinking. More on that later.

Meanwhile, back to my stomach.

Digestion

When we left it I had just eaten steak, chips and veg, along with a glass of fruit juice. The fruit juice passed through my stomach with hardly a pause, and on to my small intestine where it gave me a sugar hit. My 'second brain' or enteric system that controls digestion decided to keep the solid part of my meal in my stomach for further processing.

Thanks to the pill cam, which my stomach had obviously decided was a particularly tough bit of gristle that needed a lot of extra pounding, I could watch the whole fascinating process unfold. The solid bits of food were soon being churned up by the muscles in my stomach, while at the same time being doused in gastric juices, which are released at the sight, smell or even thought of food. It's this combination of movements and chemistry that turns lumps of food into a creamy mush known as chyme.

Chunks of green veg would fly past the camera, along with less obviously recognisable white blobby bits of what I assumed must be meat – soon to be broken down into meat fibres. The veg I'd eaten, kale, was extremely fibrous, which is why my stomach was finding it hard to process. The potato chips, on the other hand, swiftly started to break apart under the action of acid and enzymes. Reduced to a soggy mass, they were the first of the solid food I'd eaten to be passed on to the next part of the digestive process, my small intestine.

Like the sugary drink that had gone before, foods like pasta, potatoes, rice and bread are low in fibre and rapidly converted into energy. Once they reach your small intestine they are broken down and absorbed to give you a big blood sugar surge. Foods that are rich in protein, fat or fibre, on the other hand, take much longer to be processed and absorbed.

Cutting back on readily absorbable carbs, like bread, potatoes and rice, will probably help you avoid big blood sugar surges, and thus prevent weight gain and type 2 diabetes. However, as we shall see, some people are far more sensitive to these sorts of foods than others.

The green veg I had eaten hung around in my stomach for a surprisingly long time, as did the steak. It can take a couple of hours for a steak to be digested and give up its nutrients, which is one reason why we talk about having a 'second stomach' when it comes to dessert.

If your body hasn't finished digesting your main meal, your brain may still think, 'I'm hungry, I could murder a bit of cheesecake.' If you eat slowly, taking time over your meal and perhaps pausing after the main course, you will find the desire for a sugary dessert begins to fade.

There's another good reason to linger over your meal, and that is to give your hunger hormones time to kick in.

Appetite hormones

To accommodate a big meal, your stomach has to expand from the size of your fist to something more like the size of your head. That's a 40-fold increase. It used to be thought that the body regulated weight mainly via the stretch receptors in the stomach. As the stomach expands these stretch receptors send signals to the brain telling it that the stomach is full, and it's time to stop eating. But it turns out that appetite control is nowhere near as simple as that. The gut produces lots of different hunger signals, which play a far more important role in how much we eat than the stretch receptors.

Among the different hormones that regulate appetite, there are some that make you hungry and others that tell your brain when you are full.

The main hormone that makes you hungry is ghrelin.

The easiest way to remember is to think 'gre for greedy'. It is secreted by your stomach when it's empty and travels to a part of the brain called the hypothalamus. There it creates that familiar urge to eat. When you go on a diet your ghrelin levels tend to rise, which makes sticking to it that much harder.

The hormones that tell you when you are full include leptin, which is made by your fat cells. Once released, it tells your brain that you have enough fat on board, no need to eat any more. Unfortunately, when you go on a diet, cut your calories and begin to lose body fat, those fat cells respond by cutting their output of leptin, so you start to feel hungry a lot of the time.

Once upon a time it was hoped that a simple way to cure obesity would be to inject overweight people with leptin. It worked well in mice, but not on humans. The problem is that people who have been overweight for a while have often become severely leptin-resistant. They have lots of leptin running around their system shouting 'stop eating, you have plenty of body fat', but their brains are no longer listening.

Something similar happens in type 2 diabetes. If you have a diet high in sugar and simple carbohydrates, like pasta or white rice, your blood sugar levels will be constantly spiking. Your pancreas responds to this by producing the hormone insulin, whose job is to bring those sugar levels down. But, like kids who are constantly shouted at, after a while your cells stop responding to insulin.

Your pancreas pumps out more and more, but it has less

and less effect on your blood sugar levels. Instead, the insulin causes your body to go into 'fat storage' mode (that's one of insulin's other jobs), diverting extra calories away from your muscles and into your fat cells. As your insulin levels rise, you become hungrier, tired and, of course, fatter. Eventually you may end up as a type 2 diabetic on medication.

Fortunately, there are other hormones whose job it is to suppress the appetite, including one called PYY, which is produced in a number of places, including the stomach and small intestine.

PYY levels go up after you've eaten, which is helpful because PYY not only suppresses the appetite but also delays gastric emptying, which gives your guts more time to absorb the nutrients you've just eaten.

The good news is that overweight people do not seem to develop a resistance to it in the way they do with leptin. The bad news is that attempts to create a weightloss drug based on PYY have not been successful, largely because of unacceptable side effects, like nausea.

You can naturally boost your levels of PYY by eating protein, which is why I tend to eat eggs or fish for breakfast (their high protein content keeps me pleasantly full and hunger-free for far longer than if I take in exactly the same number of calories in the form of cereal or toast).

You can also boost your PYY by eating more slowly. It takes time for the food you eat to pass through your stomach, reach your small intestine and then activate those PYY cells. If you wolf down your meal, you will eat considerably more than if you eat in a more leisurely manner. If you can, eat at a table and not on the run. Try to put your knife and fork down part way through the meal and give your PYY-producing cells time to spring into action.

As I mentioned earlier, one of the depressing things about going on a diet is that as you shed fat, the levels of your hunger hormone, ghrelin, rise. At the same time, the levels of your appetite-suppressing hormones, like PPY, fall. Your body is trying to hold onto fat, sabotaging your attempts to get rid of it. This is an unfortunate hangover from a time long ago when having a thick layer of fat was like having money in the bank, the key to survival through periods when there wasn't much food about.

The fact that the body seems to be so reluctant to give up its fat has led to some pretty apocalyptic claims about dieting, such as that '95 per cent of diets fail', or 'It is impossible for most people to shed weight and keep it off'.

Commonly quoted though it is, '95 per cent' is an invented statistic, along the same lines as 'We only use 10 per cent of our brains'. I've tried to find a credible basis for this claim and it simply isn't there.

Yes, there are plenty of diets that fail, in particular the low-fat diets that have been so heavily promoted over the years. But there are others, like the Mediterranean diet, that have a very decent track record.

Long-term studies on diets based around Mediterranean eating (a way of eating I am passionate about, for all sorts of reasons – see my book, *The 8-Week Blood Sugar*

Diet) have found that, although there can be a bit of weight regain around the six-month mark, if you keep the weight off for a year, there is a good chance of keeping the weight off for many years.[2]

This is corroborated by a recent study done at the University of Copenhagen, which looked at what happens in the long run to your appetite hormones.[3]

In this study they got 20 healthy but obese people to go on an 800-calorie diet for eight weeks. The volunteers lost, on average, 13 per cent of their body weight. They were then put on a one-year programme, which included psychological and lifestyle support. Contrary to the myth of inevitable weight regain, they kept it off. They all had bloods taken to measure the levels of their appetite hormones, both before they lost weight and after a year.

The scientists found that not only did the levels of the hunger-promoting hormone, ghrelin, return to normal, but that those of the hunger-suppressing hormone PYY were 36 per cent higher at the end than at the beginning.

This study shows that if you are able to maintain weight loss for a while (and it may mean up to a year), your body will finally 'accept' this new weight as normal.

'At this point,' Dr Torekov of the University of Copenhagen points out, 'the body is no longer fighting against you, but with you, which is good news for anyone trying to lose weight.'

Gastric surgery

I would always encourage people to try and lose weight by changing what they eat, but an alternative approach that is becoming increasingly popular is gastric surgery. Shrinking the stomach will almost certainly lead to weight loss, but it is not for the faint-hearted.

I have watched only one such operation in my life, and that was enough.

I met the patient Bob, aged 34, just before he had his operation. Bob was hugely overweight and he rarely felt full, no matter how much he ate.

'I have my breakfast,' he said, 'of toast and a bowl of cornflakes. Lunchtime is normally a couple of chapatis with some chicken or vegetable curry. After lunch, if I'm still feeling hungry, I have another bowl of cornflakes or Weetabix. In the evening I get some chips, a burger and later on a sandwich. If I'm still feeling a bit peckish I buy a big packet of crisps and munch away. I just can't get full. I want to eat everything.'

Six years earlier he had had a major health scare.

'I was at home, lying down, and the next minute I had this terrible pain in my jaw and in my arm. I knew immediately that there was something wrong. I went to the hospital and they told me that I'd just had a heart attack. I was only 28 at the time, a young man – and yet I was having a heart attack. I was devastated.'

Bob went on a low-fat diet but didn't manage to lose weight. He also developed type 2 diabetes. In the hope of curing his diabetes and regaining his health, he went to see a bariatric surgeon.

There are lots of different types of weightloss surgery. The simplest is gastric banding, whereby the surgeon puts an inflatable silicone band around the top of the stomach, reducing its ability to expand. In the US this procedure is offered to people with a BMI between 30 and 40, or who have type 2 diabetes or high blood pressure.

The idea is that it cuts the amount of food you can eat in one go, giving your PYY cells time to react and tell you that you are now full. It is relatively non-invasive and it is reversible.

But Bob's surgeon felt he needed something more, so he went for a full gastric bypass, which involves the surgeon not only reducing the size of the stomach but also rerouting some of the small intestine. It is more effective but also more dangerous. Some patients get a blockage, which requires further surgery, while others suffer gastric 'dumping', resulting in nausea, rapid heartbeat, fainting and diarrhoea.

A recent review found that those who had a gastric bypass lost more weight and two-thirds were able to reverse their type 2 diabetes, compared to a third of gastric-band patients.

In the first part of Bob's operation, I watched the surgeon create a small pouch out of his stomach, reducing it from the size of his fist to the size of his little finger. The part of his stomach that produces ghrelin was separated off, to try and ensure a permanent fall in his ghrelin levels.

The second part of Bob's operation involved attaching his new, smaller stomach further down his small intestine, to a section known as the ileum. Cells in the ileum secrete PYY. With his ileum so much nearer his stomach, Bob's PYY cells would 'see' food and start to secrete the hormone less than five minutes after it had left his stomach.

Six weeks after his operation Bob had lost over three stone and his blood sugar levels were back to normal, despite his coming off medication. The main thing he found was that he no longer felt hungry all the time; even small portions would fill him up.

'I had a bite of a burger – I couldn't swallow it. I'm keeping to my diet – I'm happy with that and I can stick with that. Family's happy, I'm happy, can't wait to lose more weight.'

Summary

- Digestion is a complex process, organised and co-ordinated by your second brain, the one that runs the full length of your intestines.

- When you have a meal, fluids are the first thing to pass through your stomach and get absorbed. This is great if you've been drinking water, tea or coffee (rich in antioxidants), but terrible if what you have been drinking is sugar-laden. Then you'll get big sugar spikes, loads of calories, and feel hungry soon afterwards.

- Like sugary drinks, easily digestible carbs, such as bread, potatoes and rice, are rapidly broken down and absorbed. Being 'easily digestible' sounds good, but unless you need the energy immediately (because you are doing lots of exercise) they will also produce a blood sugar spike, followed by a crash.

- Foods that are rich in protein, fat or fibre (eggs, meat, vegetables, wholegrains) are much more slowly absorbed and will keep you fuller for longer. Eating plenty of fibre, particularly if you get it from vegetables and grains, is very important for feeding the 'good' bacteria that live in your large intestine. We will find out much more about these Old Friends, and how to keep them happy, later in the book.

2

Into the Small Bowel... and Beyond

Once you pass beyond the stomach you reach the small intestine or small bowel. To be honest, it's not that small, ranging anywhere between three and ten metres long. It's here that your gut really begins to get to work on your food, which has already been reduced by your stomach to creamy chyme – a pulpy fluid consisting of gastric juices and partly digested food. Every few seconds the surface of your small intestine convulses as a muscular, peristaltic wave passes through. These waves serve to mix the small packages of food that the stomach delivers to the intestines in a very controlled way. It's vital that they are well co-ordinated, otherwise there would be serious blockage further down.

As the tight folds on the wall of the small intestine churn the chyme in a frantic corkscrew motion, digestive enzymes from your pancreas and gall bladder begin to flood into the gut. These enzymes break your food down, in much the same way that the enzymes in washing powder break down dirt, making it easier for your body to absorb the nutrients. The bits of food that your body can't digest are sent down the gut to the ever-hungry microbes that lurk in the large bowel.

There are a few microbes in the small bowel, but there are far more further down.

The pill cam I'd swallowed missed the opportunity to film a lot of this exciting digestive frenzy because my pyloric sphincter, that picky doorman guarding the entrance to the small bowel, stubbornly refused to let it through until almost everything else had passed. This, I thought, showed a remarkable degree of intelligence (you should never allow television cameras to go just anywhere), but it was frustrating.

Finally, though, I got my first look at the inside of my small bowel. It reminded me of the last time I went snorkling in a bay full of seaweed. It was murky down there, with bits of partially digested food floating around. Unlike the relatively smooth surface of the stomach, this part of the intestine is covered in lots of villi, millions of little fingery projections that make it look a bit like a fluffy towel. Their purpose is to increase the surface area of the gut, thus maximising the area available for the absorption of nutrients.

It's often claimed that the villi increase the surface area of your gut to 'the size of a tennis court', but is it true? Well, some intrepid Swedes from the University of Gothenburg decided to find out.[4]

They took gut samples from healthy volunteers,

examined them under a microscope, counted the villi, worked out their surface area, multiplied this by the length of their small intestine and came up with a figure of… 30 square metres. Their conclusion?

'The total area of the human adult gut mucosa is not in the order of a tennis lawn, rather that of half a badminton court.'

Something to think about next time you play badminton.

Common diseases of the small bowel

Coeliac disease and gluten sensitivity

The importance of the villi becomes clear if you are unfortunate enough to develop coeliac disease. This is an autoimmune disease in which the immune system, which should be defending the body against infection, starts to attack and destroy healthy tissue, in this case the villi of the gut.

The trigger for all this destruction is gluten, or rather gliadin, one of the proteins that gluten is made of.

Gluten is found in grains, such as wheat, barley and rye, though it is also present in most processed food, including things like soy sauce.

If you've ever made bread you will know that when you mix wheat flour with water it produces a sticky mess. It's the gluten in the flour that produces the stickiness (thus glu-ten) and adds elasticity to the dough, which helps it rise when baked.

Gluten is what makes bread delicious, but in some people it also triggers an immune reaction. When you eat gluten your immune system will not only attack gliaden molecules, but also the lining of your gut. The result is inflammation.

This can lead to pain, diarrhoea, bloating and fatigue. Or, like 50 per cent of people with the condition, you may have no gut symptoms at all – you may just go along for many years feeling a bit tired and run down.

I have a niece with coeliac disease. She first started to show symptoms when she was only a few years old. She was tired, frail, off her food and generally lethargic. My anxious brother and his wife took her to see a paediatrician, who assured them that there was nothing to worry about. My sister-in-law had recently given birth to another girl, and for reasons best known to himself, the doctor decided that their older daughter's symptoms were caused by sibling rivalry. He said that no further investigations were needed. Fortunately, my brother and his wife decided to push for a second opinion, got my niece tested and discovered that she had coeliac disease.

The testing itself is quite straightforward. You have a blood test (looking for antibodies), and if that is positive you may have to have a biopsy done of your small intestine. There are lots of myths about coeliac disease. Here are a few surprising facts:

1. *It is common.*
Around 1 per cent of the population have coeliac disease, and rates have soared in recent years. Unfortunately it is often missed. According to the University of Chicago's Celiac Disease Center 'at least 3 million people in the US are living with the disease – most of them are undiagnosed'. They add that it typically takes four years for someone with symptoms to get a correct diagnosis and this delay 'dramatically increases an individual's risk of developing other autoimmune disorders, neurological problems, osteoporosis and even cancer'.

In the UK it's estimated that there are at least 500,000 people with coeliac disease who don't know they have it and who've been seeing their doctor for many years complaining about gut problems. According to Coeliac UK, the average time it takes to get diagnosed is 13 years.

2. *It often starts in middle age.*
Although we think of it as a disease of childhood, the most common age to get diagnosed is between 40 and 60. Undiagnosed coeliac disease is a relatively common cause of infertility, as well as gut problems.

3. *Even if the tests are negative, you may have a problem.*
There are lots of people who don't have coeliac disease but who may have have Non-Coeliac Gluten Sensitivity (NCGS). Symptoms include cramping,

bloating and diarrhoea. Unfortunately, it is hard to diagnose, because there are no reliable tests. You arrive at the diagnosis by excluding everything else.

How common is it? In a recent Italian study researchers tested 392 patients who were complaining of gluten-related problems.[5] They found that 26 out of the 392 (6.6 per cent) had undiagnosed coeliac disease and two (0.5 per cent) had a wheat allergy. The rest were put on a gluten-free diet and followed for two years. Those who were free of symptoms after six months – 27 people (6.9 per cent) – were considered to have NCGS. So it turned out that 14 per cent of this group, who thought they had a gluten problem, either had coeliac disease or NCGS.

But what about the rest? When they complained about gluten-related problems were they deluded? I've always been sceptical about claims that gluten affects a lot of people; after all, we have been eating wheat for thousands of years. But then I took part in a revealing experiment organised by the University of Worcester.

The experiment was done for a TV series. We recruited 60 volunteers, including some who thought they had a problem with gluten, and others who were complete sceptics. We asked them to go on a gluten-free diet for six weeks, and keep a detailed record of gut symptoms, like bloating and nausea. We also did extensive blood tests.

The cunning part of this experiment was that we provided them with gluten-free pasta to eat, which we swapped for ordinary gluten-rich pasta for a couple of the weeks. We didn't tell them exactly what we were doing because we didn't want to influence the outcome.

None of the blood tests, which were looking for things like increased inflammation, showed any change. But, and this for me was a surprise, most of our volunteers reported far more flatulence and bloating during the weeks we were slipping gluten-loaded pasta into their diet, and this stopped as soon as we took it out. Once we had revealed the results, many, including former sceptics, decided to continue on a gluten-free diet.

I'm left confused. On the one hand I am still very sceptical about 'food intolerances'. I recently sent off samples of hair and blood to a range of companies who claim to be able to detect them and they all came back with a long list of foods (eggs, wheat, milk, broccoli, etc) that apparently I have a problem with. Yet no two companies provided the same list and in some cases, where I had sent in a couple of samples under different names, they came back with different lists.

So I do think there is a lot of hokum out there. But I also think people can react to foods in ways that modern science is unable to detect.

I recently met Kathy, a young student who used to love

pizza, biscuits and toast. She was regularly ill, with bloating, but assumed that this was part of being a girl, growing up.

She also got what she describes as 'brain fog' and was frequently in a foul mood, 'which I assumed was my hormones'. Then she developed pain in the joints, which her doctor said were 'growing pains'.

When these got really bad, and the headaches came on, she googled her symptoms and then put herself on a gluten-free diet. Within days all her symptoms disappeared. She left it a few weeks, tried eating pasta, and all her symptoms came back, magnified.

She's tested negative for coeliac disease and wheat allergy, neither of which she has got (this is particularly important if you have a blood relative with coeliac disease as it has a strong genetic component), so now she's going on an exclusion diet, aiming to remain gluten-free for a good six months, before trying it again.

It won't be easy because so many foods have gluten in them. Foods that you have to avoid on a gluten-free diet include anything with wheat, rye, barley or spelt. That means bread, pasta, cereals, cake, biscuits, etc. You'll also need to lay off the beer and read labels very carefully as many processed foods have gluten in them. Even things like salad dressing often contain barley.

Frankly, eliminating these foodstuffs is likely to help you lose weight and make you feel better anyway. But be aware that a lot of foods labelled gluten-free are stuffed full of sugar and other junk to make them more palatable.

You should also be aware that bread and breakfast cereals are often fortified with iron and a range of vitamins, so if you cut them out you may want to take a multivitamin to replace them.

Lactose intolerance

Lactose intolerance is another very common digestive problem that begins in the small bowel. Lactose is a natural sugar found in milk and other dairy products.

When you are young your small intestine produces a lot of an enzyme called lactase, which breaks lactose down into glucose and galactose (both are types of sugar). These are then absorbed into your bloodstream.

Many people lose the ability to make this enzyme as they grow older, so the lactose passes through the small intestine undigested. In the large intestine the lactose is pounced on by bacteria, which will turn it into fatty acids and gases such as carbon dioxide, hydrogen and the smelly one, methane. It's these that produce common symptoms of lactose intolerance, such as flatulence and bloating.

The easiest way to find out if you are lactose-intolerant is through an exclusion diet, along the lines of the gluten-exclusion diet I mentioned above (see more on this in Chapter 7). There are also other, more exotic and expensive ways to find out, which include a hydrogen breath test and a stool acidity test.

Irritable bowel syndrome (IBS)

Irritable bowel syndrome is another incredibly common gut disease, affecting around one in five people at some point in their lives. Symptoms include stomach cramps, bloating, and either diarrhoea or constipation. After rumbling on for a while, it often gets worse when people hit their thirties.

Dave is 32 years old. He has been constipated for many years, regularly becomes bloated and has recently begun putting on weight at an alarming rate. He has also begun to experience reflux, where acid from his stomach leaks up into his gullet.

Dave was on what he calls a 'beige' diet. He had cereal or toast for breakfast, sandwiches for lunch, pasta, pizza or fish and chips in the evening. Many of his evening meals were quick, rushed, carb-heavy and eaten late at night in front of the TV.

He went to his GP, who offered him medication, which made no difference. So he went to see nutritionist Tanya Borowski. Tanya has a lot of experience working with people who have IBS, which is why I asked her to provide recipes for this book.

Tanya recommended some simple lifestyle changes, such as cutting down on the booze, taking time for lunch and not having his meals late at night. She also asked him to commit to being in bed by 10.30pm at least four nights a week. Finally, she offered him a two-stage programme which works for most of those with IBS (see page 186).

She suggested that for four weeks he remove wheat from his diet, switch to grains that are browner and less easily digestible, reduce the carbs that cause bloating and cut out all dairy. At the same time she recommended he increase the amount of colour in his diet (see page 120) and have a tablespoon of apple cider vinegar with water before each meal. She thought Dave would also benefit from having some bitter greens to get his digestive juices flowing. To improve his constipation she gave him magnesium citrate and triphala, a traditional Ayurvedic herbal treatment.

After four weeks of more or less sticking to this approach, Dave had lost 4kg and was feeling a lot better. The acid reflux had gone, as had the bloating. So it was time to reintroduce some of the foods he had cut out. He started eating cheese again, but this time the proper stuff (see page 138) and he also started on full-fat yoghurt and fermented foods (see page 140), She gave him a probiotic to take, too (see page 147).

After another four weeks, Dave was symptom-free 'and ready to rock'.

The sort of approach I've just outlined is likely to work for most people who have mild IBS, but if it doesn't or yours is more severe, you might need to try something more extreme. The current best-tested approach to treating IBS was developed at Monash University, Australia, and is known as the FODMAP diet. If you are interested, do visit their website for more information (http://www.med.monash.edu/cecs/gastro/fodmap/).

FODMAP stands for Fermentable Oligo-, Di-, Monosaccharides and Polyols. These are carbs that cause digestive problems for some people. The FODMAP approach is to cut them all out, then steadily reintroduce them. It is a complicated diet which is best done with the help of a dietician who is FODMAP trained. Visit our website, cleverguts.com, for more information on finding a FODMAP practitioner.

The Appendix

Moving on down the gut, we come to the appendix, a worm-like structure that sits just below the junction of the small intestine and the large bowel. Most people are familiar with the appendix, mainly because they know someone who's had it removed because of an infection (appendicitis).

For a long time people thought the appendix was a waste of space, something left over from our distant ancestors. Now we know that it is actually a reservoir for 'good' bacteria, ready to be released after a bad bout of, say, food poisoning.

Imagine you have eaten something dodgy and your gut has become infected by some nasty microbes. Your body will unleash a powerful immune response and also try to flush the bad guys out by flooding your gut with fluid. You will experience this as dramatic diarrhoea.

Once you've purged your guts you will want, as soon as possible, to reseed them with the 'good' bacteria, to prevent more bad guys coming in and taking over. That's where the appendix can help. It acts as the Noah's Ark of the gut world, ready to repopulate your gut when the waters subside.

Summary

- Diseases of the small bowel are common and often hard to diagnose.

- Coeliac disease affects 1 per cent of the population and is often not picked up until middle age. Studies suggest it is now four times more common than it was 50 years ago.

- Gluten intolerance is also on the rise. There is no reliable test and the only way to find out if you have it is by going on an exclusion diet.

- Irritable bowel syndrome affects one in five people. It is best treated by a change in diet.

3

The Kingdom of the Microbiome

Eight long hours after I swallowed it at the Science Museum, my pill camera had travelled just five metres. Although many of my audience had drifted away, others came along to replace them. They were rewarded, around 6 o'clock in the evening, by the sight of the camera finally escaping from my small intestine and reaching the wide open spaces of my large intestine or colon, home to the legendary microbiome.

Here, for the first time, we could see signs of the alien species that shape our lives in so many unexpected ways. I have obviously seen my biome's relatives many times before; after all, 60 per cent of faeces is made up of bacteria from the colon and who doesn't, occasionally, examine their poo? But I had never previously encountered them in their natural setting.

The microbes that live in the colon are mainly bacteria, but there are also some fungi, viruses and simple, primitive animals called protozoa. Together they form a wonderfully complicated eco system. These were the creatures that I most wanted to meet.

There are trillions of them down there, so the camera couldn't, of course, pick them out individually, though it did begin to encounter small murky fragments of what is known in the trade as 'formed matter', i.e. faeces.

Think of the microbiome as a small but incredibly complex rainforest. It throbs with life, hundreds of different competing and co-existing species. We used to think their job was pretty basic: to protect our gut from invaders; to synthesise a few vitamins like vitamin K, which the body doesn't produce itself; and to produce nasty smells while gobbling up the fibre that our bodies can't digest. Now we know they do far more than that.

As I mentioned in the introduction, we know of three other vitally important things that the microbiome does:

1. It regulates your body weight. As we'll see in a moment, the microbes can decide how much energy your body extracts from the food you eat, and much more. Can your biome make you fat? It certainly can.

2. The biome not only protects your guts from invaders but it also regulates your entire immune system. Changing the mix of bacteria in your gut can reduce the number of coughs and colds you get as well as the impact of a range of allergenic and autoimmune diseases.

3. The biome takes the bits of food our body can't digest and converts them into hormones and other chemicals. The study of how the creatures in our guts affect our brains has its own name, 'psychobiotics' and is one of the most exciting areas of current biome research.

In the next chapter I am going to cover some of the things you can do to ensure that you and your biome stay healthy. I'm also going to justify some of the claims I've just made, because on the face of it they are quite outrageous. But first, let's meet some of the Old Friends that live in your gut.

The Old Friends

Microbes don't have teeth or claws, let alone arms or legs. But they are brilliant chemists. They can take simple compounds, such as carbon or sulphur, and turn them into an extraordinary number of different chemicals. They use these skills to extract food from unlikely places (bacteria can live on rock and radioactive waste), to destroy competitors (early antibiotics, like penicillin, were created by fungi) and to manipulate us. When it comes to chemistry, the microbes that live in your gut are among the most ingenious of them all.

Although there are hundreds, possibly thousands, of different species living, breeding and dying in your gut, until recently it was impossible to identify most of them. That's because they won't grow outside the warm, airless environment of your intestines. It would be like bringing fish on to dry land and hoping they'd thrive. So, although we were all flushing billions of bacteria down the toilet, there wasn't much that scientists could do with that valuable material.

Then came the genetic revolution and soon we were able to find traces of bacteria by looking for fragments of their DNA. It's rather like DNA finger-printing, which allows the police to trace a criminal even when they are no longer present at the scene of the crime. Instead of relying on eyewitnesses, they can now collect a drop of blood from the crime scene, and with this they are able to create a genetic profile of the offender.

Scientists know which bacteria should be in your gut, and can match what they find in your poo to giant databases of likely suspects. In fact, there are now several companies, including British Gut, American Gut and uBiome, which will sequence your microbiome for under £100 and provide you with a report of the results.

The process is simple. You go to their website, log on and send them money. They send you a plastic tube, a spatula and instructions on how to collect a faecal sample. You put your faeces in the tube, give it a good old shake, stick it in the post and wait a while to get the results. If you are not living in the US, expect the whole thing to take at least two months.

I recently sent my faeces off to uBiome and the results,

when they came back, were certainly interesting. The first bit of information comes under the title, 'Body Weight', which gives you an idea of what most people who use the service are interested in.

So what did they find in my gut, and how do the contents compare with the average microbiome?

Firmicutes

One third of my gut bacteria were Firmicutes, which is lower than average. Outside the body, Firmicutes do a whole range of things, from fermenting beer and wine to cleaning up toxic waste. Inside your guts one of their main tasks is to help digest the fat in your diet. Once upon a time, when food was less plentiful, this was an incredibly useful thing because it meant people with lots of Firmicutes could extract the maximum energy from all the fatty things they ate.

These days having a lot of Firmicutes is not such a good thing, and has been linked to a higher risk of obesity. People with high levels of Firmicutes tend to be those who live on a typical Western diet (i.e. high in fat and sugar).

Bacteroidetes

Just over half (55 per cent) of my gut bacteria were

bacteroidetes. This is a good thing, according to uBiome, because having a microbiome tilted in favour of Bacteroidetes is associated with a lean body type and with less inflammation in the gut.

This group of bacteria also play an important role in teaching your immune system how to behave (they help control how violent your immune response is) and in breaking down the undigested fibre from vegetables that make their way into your colon. When they do this they produce a number of valuable substances, including butyrate.

Butyrate

From our point of view, butyrate is a wonderful chemical. It helps control the growth of our gut wall cells, and that protects us against bowel cancer. It also has powerful anti-inflammatory effects.

Inflammation is one of the ways your body defends itself. If you get the flu you will probably become hot and inflamed, with achy joints, as your body tries to destroy the viruses. That's fine. The problem comes when inflammation persists. Just as having a chronically infected tooth is a bad thing, having persistently inflamed guts can lead to pain, bloating, gas, constipation and eventually bowel cancer.

One way to counter this is to boost your butyrate levels by eating food with lots of fibre in it. The undigested bits of fibre will reach your colon and give your 'good' bacteria plenty to chew on. Well fed, these 'good' bacteria will provide you with lots of lovely butyrate.

As well as lowering inflammation, butyrate helps to maintain your gut lining, the barrier that keeps bacteria and other toxins from escaping into your blood. If this barrier starts to break down, you get a condition known as 'leaky gut syndrome' which can lead to all sorts of distressing problems, including IBS. (See page 60 for more information.)

The recipes at the back of this book contain foods proven to boost your butyrate levels.

Akkermansia

As well as lots of Bacteroidetes I was pleased to see that I had decent levels of a species called *Akkermansia*. *Akkermansia* are unusual because, unlike most of the microbes in your gut, they don't live on the remains of your food. Instead, they live on mucus, the snotty stuff that your gut wall secretes to protect itself against invading bacteria. Like Bacteroidetes, *Akkermansia* strengthen the gut wall and reduce inflammation. The more *Akkermansia* you have, the better.

We know this, in part because of a series of recent experiments done by scientists from Belgium, where *Akkermansia* was first identified.[6]

They showed that if they gave *Akkermansia* to overweight mice it stopped them becoming obese and developing diabetes. It even worked when the bacteria had been pasteurised – heated to above 70°C. Willem de Vos, one of the researchers, said, 'This came as a complete surprise. Even more surprising was the fact that the bacterium was in some ways more active after pasteurisation: not only reducing obesity and diabetes, but preventing these diseases from developing in the first place.'

A protein, with the unsexy name Amuc 110, found on the outer surface of the bacteria, seems to be responsible. Human studies are underway and a company has been created to scale up production of both *Akkermansia* bacteria and Amuc 110.

In the meantime, you can boost your own levels of *Akkermansia* by eating more polyphenol-rich foods (see the box on page 120), and also by fasting. Since *Akkermansia*, unlike most of the bacteria they have to compete with, are not dependent on the food you put in your mouth (they live, as I mentioned before, on the mucus in your gut wall), they will thrive when you cut down your calories. I recommend intermittent fasting as part of any regime to boost your gut health. Think of *Akkermansia* as your Fasting Friends. More on this in Chapter 6.

Christensenella

Sadly, what the uBiome people didn't find in my gut was a lovely little bacterium called *Christensenella*. People who have lots of them tend to be lean, despite sometimes rather poor diets.

Zoe Budd is tall and slim. Her faeces were recently examined and found to have 13 times the normal levels of *Christensenella.*

'I have a pretty terrible diet,' she told me. 'I love croissants, bacon sandwiches, cookies and cakes.' She does try to eat plenty of vegetables but always goes for the pudding and the chocolate cake when given half a chance.

Christensenella seems to be one of those bacteria you inherit from your mother and which only thrive in people with the right genes. It's not clear, if you don't already have them, how you would go about acquiring them. There is no evidence, for example, that consuming them in capsule form will do you any good (although you can buy them on the internet).

Zoe's kids are like her, tall and slim, so maybe they've inherited a few *Christensenella* bacteria, along with half her genes.

Lactobacillus

I was also surprisingly low in *Lactobacillus.* This bacterium lines your intestines and is good at protecting your guts from bad guys, such as the fungal pathogen *Candida albicans.* Many women will be familiar with this nasty little beast because it is responsible for vaginal thrush. *Lactobacillus* destroys pathogens like *Candida* by spraying or injecting them with hydrogen peroxide, the stuff that people use to bleach their hair.

If your *Lactobacillus* are weakened by a course of antibiotics that you have taken to treat something else, this can give *Candida* a chance to take over in your gut. Once it does it is hard to dislodge.

Some strains of *Lactobacillus* are important for mental health; at least those who take it in the form of a probiotic (a capsule containing live bacteria) report that it improves anxiety and mood.

In combination with other probiotics it may also help people with IBS. More on that later.

Bifidobacterium

Finally, one bacterium that I have lots of, in fact three times more than average, is *Bifidobacterium.* These guys break down otherwise indigestible fibre and, like *Lactobacillus*, protect your gut from other less friendly microbes.

You get them early in life from your mother's breast milk, and you will also find lots of them in cheese and yoghurt, two of my favourite foods, which could explain why I have so many in my gut. Like *Lactobacillus,* they are a popular probiotic.

Leaky gut syndrome

As well as protecting your gut from attack by unfriendly bacteria, 'good' bacteria help reinforce your gut wall, making sure that bad stuff doesn't get out of your gut and into the rest of your body. When that happens you get 'leaky gut syndrome'.

If you tell a doctor you have leaky gut syndrome they will probably roll their eyes and sigh. Many doctors think it is an imaginary condition dreamt up by loopy promoters of alternative medicine to flog dubious supplements to desperate customers. It's certainly what I used to believe.

Yet there is now clear evidence that leaky gut, or 'intestinal hyperpermeability' is a real condition. It is characterised by bloating, pain, gas, cramps, tiredness, food sensitivities, as well as generalised aches and pains.

It occurs following a gut infection, a course of antibiotics, a poor diet or even a strenuous bout of exercise, when the tight junctions between the cells that form your gut lining start to loosen. Think of the lining as the Great Wall of China, designed to keep invaders out, but which now has some bits missing, through which the enemy can squeeze. Once bacteria and other toxins get out of your gut and into your blood you are in trouble.

Diagnosing leaky gut

A leaky gut can be hard to diagnose, which is why many doctors are so sceptical about its existence. One way is to use a 'sugar probe'; this involves giving you a mix of two different types of sugar to drink. One of these sugars must be small and simple, chemically speaking (mannitol), while the other should be bigger and more complex (often lactulose).

In a normal gut the small sugar molecules will be able to squeeze between the gaps in the gut wall, but the large ones won't. If you have a leaky gut even the large molecules will get through, and a much higher ratio of lactulose to mannitol will be detected in your urine.

Recent studies, using sugar probes, have shown that common gut disorders, like IBS, which are often dismissed as being 'all in the mind', frequently do have a physical cause.

Understandably, people who have spent years with pain and bloating become anxious and depressed. But antidepressants may not sort them out. A change in diet, to one that particularly encourages the growth of 'good' bacteria, just might. See the recipe section at the back of the book for some excellent, soothing broths and smoothies to help repair the gut lining.

The bad guys

As well as microbes that keep your gut healthy, there are plenty of 'bad' guys on the prowl, microbes that can cause serious gut problems, given half a chance. I've done a lot of travelling, particularly in some of the poorer parts of the world, so I am personally familiar with quite a few of these.

Campylobacter

This is the commonest cause of food poisoning in affluent countries. Symptoms include abdominal pain, severe diarrhoea, sometimes with blood, and vomiting. It can lead to IBS and arthritis. It can also cause pregnant women to miscarry.

You typically pick up *Campylobacter* by handling raw chicken or undercooking it. It takes a single drop of juice from a raw chicken to infect someone, and the bacterium is present in the majority of chickens sold in supermarkets.

Washing a chicken before you put it in the oven is a terrible idea because it just spreads the bacteria around. Instead, just make sure you cook it properly.

While most people have heard of *E. coli*, *Salmonella* and *Listeria*, *Campylobacter* causes more cases of severe food poisoning than those three put together.

E. coli

Many types of *E. coli* are harmless, but some can cause bloody diarrhoea leading to anaemia (severe blood loss) and even kidney failure. They are also a common cause of urinary tract infections. You commonly get a gut *E. coli* infection from eating infected meat that has not been fully cooked (i.e. to at least 71°C). You can also pick it up from drinking raw milk.

Salmonella

Another little delight is *Salmonella*, the food-poisoning version of which causes diarrhoea, stomach cramps and vomiting. Typically symptoms begin within 12-72 hours of swallowing it and lasts up to a week. If you become seriously ill you may need to go to hospital, because you could die from dehydration. *Salmonella typhi*, which causes typhoid fever, is much nastier and more likely to be fatal. It infects over 20 million people, worldwide, every year.

Clostridium difficile (also known as *C. diff*)

This is one of those bacteria that may well be hanging around in your guts right now, doing little harm, being kept in its place by your 'good' bacteria. Then you get a minor infection and your doctor prescribes you a broad-spectrum antibiotic. This wipes out much of your normal gut biome, but leaves *C. diff* intact, because these are tough and normally very antibiotic-resistant bacteria. With its rivals destroyed, *C. diff* may seize the opportunity to take over. It is as a bit like the invasion of Iraq: once the allies had destroyed Saddam, they created the circumstances in which ISIS could thrive.

Once *C. diff* are in charge they are incredibly hard to get rid of. Symptoms include watery diarrhoea, fever and abdominal pain, which can go on for many years. Incredibly, *C. diff* kills some 29,000 people every year in the US and costs around $1.5 billion a year to treat.

Standard treatments, which include yet more antibiotics, are often ineffectual. In which case your best bet may be something that sounds rather unpleasant: a faecal microbial transplant or FMT.

Faecal Microbial Transplant (FMT)

As the name implies, FMT involves doctors taking faeces from a healthy donor and squirting it into an infected patient. The hope is that the microbes from the healthy donor will overwhelm and destroy the *C. diff* bacteria. The transplant can be done via a naso-gastric tube, but there is a higher success rate if it is done via a colonoscopy and then squirting in the donor poo. The main

downside, apart from being distasteful, is that the donor poo has to be properly screened to ensure it's not harbouring anything nasty. I have seen one FMT done and the whole thing took less than 30 minutes. The results were spectacular.

Rose had spent many years in pain, having to rush to the bathroom at all hours of day and night. There had been countless failed attempts to treat her with antibiotics. Yet she was cured within hours of being given a faecal transplant and was able to return to normal life shortly afterwards.

Although it's relatively new, FMT is amazingly successful. Randomised controlled trials have shown that a staggering 94 per cent of patients make a full recovery after a single faecal transplant.

The treatment is now being tried out on all sorts of other gut conditions, from IBS to leaky gut syndrome and type 2 diabetes. People have even started doing DIY versions, but this is one procedure I really wouldn't try at home.

In search of diversity

It's not just the type of bacteria in your gut that matters, it's also the diversity. You want your biome to be as multicultural, as possible. The loss of a minor species won't have too much impact; but if a few 'bad' species, like *C. diff,* begin to take over, the effect can be disastrous. A diverse microbiome will allow your gut to recover much more quickly from a bout of diarrhoea.

Unfortunately, like animals in the wild, many species in our gut are in decline and have been for decades. It's partly because we eat such a narrow range of foods, which means our gut bacteria also have to live on a restricted diet. Of the 250,000 known edible plant species, we use less than 200. Seventy-five per cent of the world's food comes from just 12 plants and five animal species. It's one reason why I am going to encourage you to branch out a bit and try things like fermented foods, which you may never have considered before.

Another reason for the decline is the widespread use of antibiotics, not only to treat us but also to help the animals we eat put on weight. Routine use of antibiotics in animals as a growth promoter is now banned in Europe but is still common practice in other parts of the world, including the US.

Finally, there are emulsifiers. These are chemicals that are added to processed foods to extend their shelf life. They've been shown to reduce microbial richness, and may directly contribute to colitis and diabetes.

How to make your biome more diverse

There is a measure, called the Simpson index, which tells you just how diverse your biome really is. My most recent score was 7.99, which puts me into the top 30 per cent of 'most diverse microbiomes' among those who have sent poo samples to uBiome. This, however, is not much to brag about. People like the Hadza (a tribe of hunter-gatherers who live in Tanzania) have a biome which is at least twice as diverse as mine.

A couple of years ago Jeff Leach decided to try and acquire a healthier gut and blog about it under the title 'Going Feral: my one-year journey to acquire the healthiest gut microbiome in the world (you heard me!)'.

His attempts to acquire a healthier gut included hanging upside down under a tree in Africa while someone poured liquidised poo from a Hadza tribesman into his bum via a funnel.

Jeff is not particularly wild or deranged. In fact, he is a very well-respected scientist and co-founder of American Gut. Based at the Knight Lab at the University of California, American Gut describe themselves as 'one of the largest crowd-sourced, citizen science projects in the country'. Like uBiome, for a modest donation, they will take a look at your poo and let you know what is in it.

Things they have discovered so far include:

- The more different types of plants you eat, the more diverse your microbiome.
- People who have at least one alcoholic drink a week have a more diverse microbiome than those who don't drink at all.

In his search for 'The Healthiest Microbiome in the World', Jeff was not about to confine himself to eating a few more plants and having the occasional drink. He decided to move in with the Hadza, adopt their diet, 'eating the plants, animals and drinking the same water (with the occasional baboon turd floating in it)', and see what happened.

So what did happen? Well, the initial impact on his gut was a decrease in microbe diversity, although it soon bounced back.

He also saw a big increase in our friend *Akkermansia*. As I've mentioned before, *Akke*r is good because high levels of it are associated with low levels of obesity. It also protects you from invasion by nasty microbes by reinforcing your bowel wall. In the weeks he was with the Hadza his *Akkermansia* went from almost non-existent to being a dominant player.

He thinks there are a number of reasons for this, including the fact that he, like the Hadza, went for quite long periods without food. As he writes in his blog:

'I'm gonna go out on a limb here and suggest that the Hadza – and presumably our ancestors – were periodically fasting… I don't mean going all day without food – there's plenty of food in Hadza land. Plenty. But rather they don't immediately solve hunger pangs by opening a bag of chips as it takes just a little bit of effort to get some food and sometimes you delay your desire to eat because of it. This little bit of extra time between mouthfuls is what I mean by intermittent fasting in the case of the Hadza and what I experienced when I was following the diet.'

On his blog, humanfoodproject.com, Jeff makes a number of suggestions about what you can do to ensure a healthier biome, in addition to cutting down on the snacks. These include:

1. Avoid antibiotics if you can, because a course of broad-spectrum antibiotics can 'take weeks, months or even years for your gut microbial community to bounce back from – if at all.'

2. Open a window. We spend 90 per cent of our lives living indoors. Studies show that opening a window increases the diversity of microbes in your house and therefore, presumably, in you.

3. Eat more plants. That doesn't necessarily mean giving up meat, nor does it mean simply eating a lot more carrots. It means eating as wide a range of different plants as possible. But it also means eating lots of different bits of the plant, not just the tasty bits. 'Consume the entire asparagus, not just the tip. Consume the trunk of the broccoli, not just the crown.'

4. Get your hands dirty. Preferably by gardening. This connects you with the trillions of bacteria that live in the soil and is a good way of getting in some exercise.

The end of the pill cam's journey

We met Pill Cam, the camera that I swallowed in the Science Museum, at the beginning of this chapter, just as it entered the kingdom of the microbiome. I got a brief glimpse of that gloomy realm, but the further the camera went, the murkier the images became. Anyway, by then it was late and the museum was keen to close. So, with little prospect of any more useable footage being beamed back from this strange and alien world, we decided it was time to end the show. The audience went home, and so did I.

Unobserved, Pill Cam continued on its merry way, along my large bowel, all 1.5 metres of it, through my muscular sigmoid colon and into my rectum. Finally, many hours later (about 12 hours, if I remember rightly), it exited my body. You'll be pleased to hear I've cleaned the camera up, polished it and kept it as a memento of a most unusual day.

In all, it took the pill cam around 20 hours to travel from my lips to the toilet bowel. This is pretty fast going, as a typical transit time (the time it takes from eating to excretion) is, according to the Mayo Clinic, normally more like 40 hours. In one study they did, with healthy volunteers, they found the transit time in men averaged 33 hours, while women took a more leisurely 47 hours.[7]

If you are a keen self-experimenter and want to time yourself you could try eating food with red food dye in it and see how long it takes for the dye to appear in your faeces.

Summary

- A healthy biome is home to a rich diversity of microbes. It is a delicately balanced ecosystem where bacteria vie for supremacy.

- Some, such as *Lactobacillus* and *Bifidobacterium*, produce substances like butyrate that damp down inflammation. They also protect your gut from attack by unfriendly microbes and help reinforce the walls of your gut, preventing leaky gut syndrome.

- Although too much junk food and overuse of antibiotics may have damaged your biome, you can shift it in a healthier direction by changing what you eat, what you drink, and by embracing the great outdoors.

- These days you can get your biome sequenced for less than £100. This will tell you what's down there and what shape it's in. If you repeat the tests after following some of the advice in this book you should soon be able to tell if the Clever Guts Diet has made a difference. Although you will also be able to tell from the scales and how you feel.

4

How your Biome Influences You, and How You Can Influence It

As we have discovered, a healthy microbiome should be made up of as rich and varied a range of microbes as possible. There are trillions of them down there and together they add up to roughly the same number of cells as the rest of your body combined (microbes are generally much smaller than human cells).

As I will show you in a moment, they affect things like your mood, your weight and how active your immune system is. While they influence you, you can also change them. What you eat and the way you live will have a huge impact on the sort of creatures that thrive in your internal rainforest.

But how do you acquire these microbes in the first place?

Where your microbiome comes from

Most of us get our biome from our mothers. When you were

in her womb your gut was largely free of microbes. What happened next depends on the way you were born.

If you were born vaginally, you would have travelled down the birth canal and taken a good gulp of your mother's faeces and vaginal fluids. With no competition from existing microbes, these bacteria would then have rapidly colonised your guts. Later, if you were breastfed, you would also have acquired 'good' bacteria from her breast milk.

If you were born by a Caesarean section (C-section), things would have been very different. The first bacteria you encountered would have been those hanging around in the operating theatre, those on the skin of the people that held you and anything else you inhaled or swallowed in the first few hours of life.

Poo samples taken from babies who were delivered by Caesarean section show that they have a very different gut bacterial population from those who were born vaginally.

Does this matter? Yes, very much so. Thanks to a large study done by researchers from the Harvard T.H. Chan School of Public Health,[8] we know that babies who are born by Caesarean section are far more likely to become obese children and overweight adults.

In this particular study, aptly known as GUTS (Growing Up Today Study), researchers tracked more than 22,000 babies from birth into adulthood. They weighed the children when they were teenagers, and again when they were in their twenties.

The results were dramatic, particularly when they compared brothers and sisters. Children born via C-section were 64 per cent more likely to be obese than their siblings born vaginally, despite sharing the same genes, being brought up in the same household and eating the same food. This strongly suggests that it is the gut bacteria that are making the difference.

'Caesarean deliveries are without a doubt a necessary and lifesaving procedure in many cases,' said Jorge Chavarro, senior author of the study. 'But Caesareans also have some known risks to the mother and the newborn. Our findings show that risk of obesity is another factor to consider.'

Caesarean-born babies are also far more likely to develop allergies later in life.

With Caesarean rates currently running at around 26 per cent in the UK, 35 per cent in the US, 50 per cent in China and up to 80 per cent in Brazil, this is storing up problems for the future.

One way round this problem, if a C-section has to be done, is to try and expose the baby to some of its mother's microbes as quickly as possible. It's known as 'vaginal seeding' or microbirthing, and involves taking a swab from the mother's vagina and anus, and wiping it over the baby's mouth, eyes and face shortly after birth.

This approach has its fans, but Dr Aubrey Cunnington from the Department of Medicine at Imperial College London isn't one of them. He thinks that unless it is done in a carefully controlled way, the potential risks, such as picking up an infection from mum, outweigh any likely advantages.

'One colleague had to intervene when a mother with genital herpes, who had undergone a Caesarean section, was about to undertake this process,' he says. 'Swabbing could have transferred the herpes virus to the baby.'

He also thinks there are better ways to ensure your baby gets the best start in life. 'Encouraging breastfeeding and avoiding unnecessary antibiotics may be more important to a baby's gut bacteria than worrying about transferring vaginal fluid on a swab.'

Breastfeeding is certainly going to give a baby's biome a good start and may well contribute to the fact that breast-fed babies are half as likely to develop eczema or asthma as a baby who is exclusively bottle fed.

As well as a good mix of fats, proteins and carbohydrates, breast milk contains complex sugars which, oddly enough, babies can't digest. It seems that their main purpose is to feed the 'good' bacteria which are, hopefully, growing in the baby's gut. The fact that breast milk evolved to feed these Old Friends shows just how important they are.

As we get older our biome changes and the number of species colonising our gut increases, up from about 100 species in an infant to around 1000 in an adult. By the time we are three years old, our biome will be largely settled, though it can still change in response to infection, antibiotic treatment and changes in diet.

Although you can't do anything about the way you were born or how your mother fed you, you can do something about how you currently feed yourself. And your microbiome. Before we come to that, I'd like to dive into some of the many surprising ways in which your microbiome affects you as an adult.

Summary

- Your microbiome is formed in the first couple of years of life and heavily influenced by how you were born.

- Breast milk contains a healthy mix of proteins, fats and sugars. It also contains compounds that a baby can't digest, but which encourage the growth of 'good' bacteria. Oddly enough, breast milk contains substances whose only purpose is to feed Old Friends.

How your biome makes you fat

As we've just seen, being born by C-section has a lasting effect on your gut bacteria and your waistline.

Being exposed to repeated courses of antibiotics when you are a child also increases the risk that you'll become overweight, particularly if those antibiotics are given in the first six months of life.

As an adult there are various ways your particular mix of bacteria may help you put on weight:

1. Some gut bacteria extract more energy from the food you eat than others.

2. They can influence how much your blood sugar levels rise when you eat.

3. They may affect your mood and the food choices you make.

Extracting energy from food

Our digestive systems are not 100 per cent efficient. Some of the calories we take in will also get excreted. And, although much of the energy in the food we eat is extracted by our gut, we also depend on our gut bacteria to do some digesting for us. Mice with sterile, bacteria-free guts are much skinnier than normal mice, despite consuming exactly the same calories.

We also know that some bacteria, like Firmicutes, are better at extracting energy from the food we eat than others. Which means that if I have more Firmicutes in my gut than my wife (which I do), I will probably be excreting fewer calories, after eating exactly the same meal, than she does.

Evidence for this comes from a study done by researchers from the National Food Institute, who collected faeces from 32 children, half of whom were overweight and half whom weren't, and gave the samples to mice.

The mice who got poo from the overweight kids gained significantly more weight than those who got it from the slim kids, despite eating the same food. The researchers also measured the unspent energy in the mouse poo and found that those who had put on the least weight were excreting the most calories.[9]

Why weight loss is not simply a matter of eating less and exercising more

What studies like this have done is undermine the idea that has dominated weight control thinking for the last half-century: 'CICO' ('calories in, calories out'). According to CICO the reason we're fat is because we eat more calories than we burn off. The answer to obesity is eat less and exercise more.

Except, of course, it's not as simple as that. As we have just seen, the bacteria in your gut can influence how many of the calories that go in, stay in. We have also seen that the gut does not treat all foods exactly the same. Eating cake will not have the same effect on your body as eating steak.

Unlike the steak, the sugar and flour in the cake will be swiftly absorbed, making your blood sugar levels soar. In response to this your pancreas will pump out insulin to bring them down. It will also switch your body into fat storage mode. The energy from the cake will now be stored as fat.

But the surge in insulin is normally followed by a crash in blood sugar levels, which will leave you tired and hungry. Studies show that when this happens you eat more.

So you put on weight. That's why it's important to avoid frequent blood sugar spikes.

To do that you need to cut back not just on sugary drinks and snacks, but also on any foods that make your blood sugars surge.

Which ones are these? Well, there is a measure, known as the Glycaemic Index (GI), which ranks foods according to their impact on blood sugar levels.

Easily digestible carbs like white bread, rice, potatoes and pasta have a high GI score, while foods with fibre in them, like broccoli or wholegrains, score low. To see a list go to glycaemicindex.com.

Although it is a helpful, and sometimes surprising, guide to what causes blood sugar spikes (who would have thought that a boiled potato would have the same impact as eating a tablespoon of sugar?), the GI is also limited. To create a GI figure for any particular food, 10-20 volunteers are fed that food and researchers then measure what happens to their blood sugar levels. On the basis of this, they calculate and publish an average GI score.

The problem is that people respond differently to exactly the same foods. The reason? The different mix of bacteria in their guts.

We know this because of a massive study by the Weizmann Institute in Israel, the largest of its kind ever undertaken.

Personalised nutrition

The Personalised Nutrition Study done at the Weizmann Institute is one of the most fascinating studies I have come across, promising, as it does, to create the first truly personalised diets.

For this study, immunologist Dr Eran Elinav and computer scientist Dr Eran Segal recruited 800 volunteers and ran extensive tests on them. At the start of the study they took blood and poo samples and lots of physical measurements, like height and weight.

The volunteers were also fitted with skin-mounted monitors, which measure blood sugar levels almost continuously. They wore these for a week, while keeping detailed records of everything they ate and drank, as well as how much sleep they were getting and how much exercise they were doing. The Erans wanted to measure, with great precision, the impact their normal daily diet and lifestyle was having on their blood sugar levels.

All in all the scientists collected over 1.5 million blood sugar measurements.

They found, to everyone's surprise, that eating exactly the same foods had very different and often unexpected impacts on people's blood sugar levels. Some volunteers' blood sugar levels soared in response to rice, while for others it had very little impact.

Is chocolate bad for your blood sugar levels? Yes for some, but not for others.

The team used data from the poo samples they'd

collected, along with details such as age and gender, to create an algorithm, a computer model that could predict which foods would cause an individual's blood sugar levels to spike.

They then tested their algorithm on a fresh batch of volunteers, comparing their predictions with how people actually responded. So how did it do? Extremely well.[10]

Finally, they recruited yet another group of volunteers, took poo samples, and using this data created a personalised diet sheet containing 'good' and 'bad' foods for each one of them.

The volunteers were asked to follow their personal diet sheet for two weeks. For one week they had to eat foods the algorithm had predicted would have a bad effect on their blood sugar levels. For the other week they ate only the 'good' foods on their list.

It wasn't just salad and vegetables on the 'good' list and cake on the 'bad'. Some people were free to eat chocolate and drink alcohol, while others had to shun apparently healthy food.

According to Eran Segal it resulted in impressive differences. 'When they were on the "bad" diets their blood glucose levels rose up, well into the abnormal range, but when they were on the "good" diet they stayed within the healthy range.'

Curious, a friend of mine, Dr Saleyha Ahsan, went to Israel to try it out.

Saleyha's journey

Saleyha is a busy A&E doctor who is constantly on the go and often doesn't have the time to eat well. She used to be in the army, was incredibly fit, but since damaging her knee she has had to restrict the amount of exercise she does. As a result she has put on a lot of weight, which she was keen to do something about. As she told me, 'I've always watched what I eat, but right now I just can't seem to shift the weight. I need help.'

For Saleyha it is not just about wanting to look a bit slimmer. She has a family history of type 2 diabetes and knows she is at significant risk of going down that road. She also has polycystic ovary syndrome, which would be helped by losing some fat around her waist.

So she went and spent time with the Erans at the Weizmann Institute. Like all the other volunteers, she was fitted with a glucometer, under her skin, so her blood sugar levels could be constantly measured. Her sleep and activity levels were monitored by a wristband, and she was given an app to record her mood and everything that she ate. They also collected poo samples from her, which showed she had high levels of the sort of bacteria linked to obesity and diabetes.

Then she embarked on six days of eating food, provided by the institute, that would test her blood sugar responses.

Finally she was presented with two lists of foods. A green 'good' list, which included things like avocados, bananas and dark chocolate.; and a red 'bad list', which unfortunately included some of her favourite foods, such as sushi and grapes, which she'd always seen as healthy.

She agreed to eat only the green-listed foods for a month. After three weeks she started to notice a difference. She no longer felt tired during the day and her hunger pangs had gone. Her skin looked better and she had lost half a stone.

'I used to blame myself,' she says, 'for being lazy but now I realise it's not as simple as that. Once I started eating foods that produce less blood sugar spikes I stopped having ups and downs or hunger pangs.'

By changing her diet she had also changed the mix of bacteria in her gut. At the end of the month there was a marked fall in the bacteria associated with obesity and diabetes.

It is impractical for most of us to hang around in Israel with glucose monitors trying out lots of different foods to see what happens. So the Erans have created a company to commercialise their work, which is just being launched.

You register, pay, and they send you a kit which you use to collect a stool sample. You also have to fill in a short questionnaire and do an HbA1c test (this is a blood test that gives an accurate measure of what your blood sugar levels have been for the last month or so – your GP can do it).

You send these off, they analyse it and then use their algorithm to produce a personalised nutrition app that recommends specific foods/meals you should eat if you want to keep your blood sugars down. If you stick to the green foods and shun the red ones they are confident that, like Saleyha, you will lose weight and feel more in control. They see this as the first step towards a truly personalised diet. Early days, but really interesting. To find out more, visit our website cleverguts.com.

To evaluate just how effective this approach is, the two Erans have just started another trial in which they will follow a large group of people for a year. The people they've recruited for the trial have prediabetes, which means they are well on the way to developing type 2 diabetes, so it will be an important test of their ideas.

They are certainly not short of volunteers. Previous participants were so enthused by their experience that they have urged friends and family to join in.

Yo-yo dieting

The two Erans are also researching the effect of gut bacteria on yo-yo dieting. As anyone who has ever dieted knows, the big challenge is not so much losing weight as keeping it off. Many people take up a diet, shed loads of fat, plateau, despair, and pile the weight back on.

As we've seen, what makes it particularly hard to keep weight off is that, when you shed fat, your body fights back, using your appetite hormones against you. As your fat cells shrink your body produces more of the hormones that make you hungry and fewer of the ones that suppress appetite. That's the bad news. The good news is that if you tough it out those appetite hormones will reset.

Intriguingly, your microbiome also seems to contribute to the problem of yo-yo dieting.

To find out what's really going on, a team led by the two Erans took mice and fed them till they were really fat. Then the mice were put on a low-calorie diet till they'd reverted to their previous slim selves. The scientists did this several times. At the end the mice looked the same as when they had started.

But they weren't. Given the chance to eat freely, the mice piled on weight, laying down fat faster than they had the first time they had been fed a high-calorie diet. What had happened was that their biome had changed. The new mix of microbes included many that now encouraged weight gain.

'It seems that the microbiome retains a "memory" of previous obesity,' says Eran Elinav. 'This new microbiome accelerated the regaining of weight when the mice were put back on a high-calorie diet.'

Fortunately, they found a way to reverse this. And that was by feeding the mice flavonoids.

Flavonoids are made by plants to protect themselves against parasites and harsh weather. You'll find them in blueberries, cherries, blackberries, plums, grapes, tomatoes and green tea. They are powerful antioxidants – and they also encourage your body to burn fat. More on them in Chapter 5.

The Erans showed that yo-yo dieting, at least in mice, creates an unhealthy biome, rich in bacteria that are hell bent on destroying flavonoids, which in turn leads to rapid weight gain.

When they gave the mice some flavonoids in their drinking water, they discovered that the mice had 'reset'. They no longer showed accelerated weight gain when put on a high-calorie diet.

The flavonoids the Erans used were apigenin, which is found in parsley, celeriac and camomile tea, and naringenin, present in grapefruit, oranges and tomato skin. It is unlikely that eating those foods will give you as big a dose as the mice got, and it's not even clear if the flavonoids that work for mice will work for us. But they are good foods to eat, so what have you got to lose?

Summary

- Your biome helps determine how much energy you extract from food and, to some extent, how much weight you put on.

- Your biome also influences how much your blood sugar levels respond to particular foods.

- By analysing stool samples it's possible to predict which foods will make your blood sugars surge. This offers the possibility of a truly personalised diet, based on valid science.

- Eating flavonoids or foods rich in flavonoids may help you avoid regaining weight after a diet.

- More disturbingly, there is evidence that your biome can influence the food choices you make.

The biome and your brain

We like to think that we are in charge of the decisions that we make, from what we decide to eat to where we go on holiday. But the truth is we make a lot of decisions subconsciously, guided by signals and cues that we are not even aware of.

Why did you eat that muffin on the way to work? Was it because you really, really wanted it, weighed up the benefits and the costs, and decided to buy it? Or did you see it when you went into the coffee shop and just pick it up on a whim? Most of us eat mindlessly most of the time, guided by habit and influenced by advertising.

But what about our gut bacteria? Is it plausible that the tiny single-celled creatures lurking in our gut can also influence our decision-making?

I think so, and so do an increasing number of scientists who work in this area. Our microbes certainly have the opportunity, the motive and the tools to manipulate us.

I've talked before about the second brain in your gut, the enteric system that contains the same number of neurons as you'll find in the brain of a cat (I love that fact!) and which communicates with the brain in your head via the vagus nerve. The vagus nerve is like a busy telephone line, with lots of messages going in both directions. The enteric system talks to your brain and your brain talks back.

There is plenty of evidence that your microbes can hack into this system and talk directly to your brain via the vagus nerve. They also produce a range of hormones and neurotransmitters that reach your brain via your bloodstream.

Dopamine, for example, is well known as a 'feel-good hormone'. The microbes in your gut produce lots of it, possibly to reward you for doing what they want you to do (like eat more cake). The microbes in your gut also produce chemicals that control your mood, like serotonin and GABA (a neurotransmitter than acts in a similar way to the anti-anxiety drug, valium). They even make a range of chemicals that are strikingly similar to leptin, ghrelin and PYY, the main hunger hormones.

So microbes have means. They also have the motive. Life in your gut is not for the faint-hearted. It is a state of endless war. Life down there is nasty, brutish and short. The microbes are not only competing for space and scarce resources; they also have different dietary demands. Some thrive on sugar, others love fat. The more sugar you feed the sugar eaters, the more they want. They are not like a friendly dog, hanging around waiting gratefully for whatever comes their way. They are fighting for life. They will do anything to give themselves the edge.

'Microbes have the capacity to manipulate behaviour

and mood through altering the neural signals in the vagus nerve, changing taste receptors, producing toxins to make us feel bad and releasing chemical rewards to make us feel good,' says Dr Aktipis, from the Arizona State University Department of Psychology. She has recently collaborated on a major review of the scientific literature, which asked the question: 'Is eating behaviour manipulated by the gastrointestinal microbiota?' The answer was a resounding 'yes'. They put together a convincing case that microbes not only influence how much we eat but what we eat.[11]

They point to research that shows that having a wide range of different species in your gut normally means you are slimmer and healthier. Having a more limited ecosystem is associated with being overweight and sickly.

Why should this be? Well, the argument goes that in a diverse microbiome all those tiny creatures will clamour to be heard and, like a gang of children shouting at once, they can be ignored. The problem comes when one group – say, the ones who thrive on junk food – start to dominate. As a gang, these bad guys will now be much more influential and, by producing chemical signals, generate cravings for junk food, cravings that you will find it hard to resist.

So how sure are we that this actually goes on? At the moment the most convincing evidence that microbes can alter behaviour comes from animal studies. It has been known for a long time that mice, when infected by bacteria called *Toxoplasma gondii*, become strangely reckless. Normally a mouse will keep to the shadows and avoid anything to do with cats. But when they are infected by *Toxoplasma* there is a complete change of behaviour. They are now attracted by cat urine and will deliberately move out into the open, to be killed. The *Toxoplasma* bacteria are manipulating the mouse's behaviour. They want it to be eaten by a cat, so that they can then infect the cat.

Similarly, if you feed faeces from mice who display anxiety-like behaviour (no one talks about 'anxious mice' because we don't know what they are thinking) to mice who don't have *Toxoplasma* in their gut, then those mice start to behave in a more anxious manner.

Scientists have found they can make mice swim for longer, and more doggedly, if they feed them bacteria called *L. rhamnosus.*[12] Further more, rather brutally, they found that if you sever the vagus nerve in these mice, so that the gut can no longer talk to the brain, the mice do not change their behaviour after being fed these bacteria.

When it comes to humans, the evidence is more circumstantial. In one study involving 22 people,[13] half of them chocaholics and the other half indifferent to chocolate, the researchers found that those who loved chocolate had a very different set of microbes from those who weren't interested in the sweet stuff, despite the fact that both groups had been put on an identical diet.

There was also an impressive double-blind, randomised, placebo-controlled trial in which researchers showed that they could significantly improve depressed people's moods by giving them a probiotic containing a mix of *Lactobacillus acidophilus, Lactobacillus casei* and *Bifidobacterium bifidum.*[14]

Turning their attention to infants, the ever curious scientists found, when they looked into their nappies, that children with colic (inconsolable crying) had reduced gut diversity and fewer bacteroidetes (the 'good' guys) than children who didn't cry as much. One theory is that lots of screaming means parents are more likely to pay attention, fuss and feed, thereby providing the microbes, which are irritating the baby's guts and causing it to scream, with more food.

A rather spooky suggestion that emerges from all this is that cravings could be catching, jumping from one person to another.[15] Perhaps cravings, and the resulting tendency to put on weight, are not just socially contagious but literally infectious, like a cold?

To be honest I find this a bit far-fetched. I have lived with my wife for more than 30 years and she has shown no signs of being infected by my sweet tooth.

Along with the rather frightening possibilities that this research throws up, there are some consoling thoughts. If cravings really are being generated by a group of cunning bacteria in your gut, then you have it in your power to starve them out.

All you have to do (and I don't underestimate how tough this can be) is keep away from the thing you crave and hopefully the cravings will reduce as those microbes die off. There will be more on this subject in the next chapter.

Summary

- The microbes in your gut can communicate directly with your brain via the vagus nerve.
- They also produce hunger hormones and neuro-transmitters which they may use to influence our cravings and our behaviour.
- If so, then changing the microbes in your gut may also change your cravings.

Allergies and gut bacteria

Two hundred years ago, life expectancy in the developed world was less than half what it is today. People died young of infectious diseases like typhoid, cholera and tuberculosis. What they didn't commonly have were autoimmune diseases, like type 1 diabetes, or allergies, like asthma. None of the characters in Jane Austen's 19th-century novels ever complains about hay fever or having a food intolerance. Charles Dickens never suggested that Oliver Twist or any of his artful dodger friends suffered from eczema. Allergic diseases are a modern plague, a product of the 21st-century lifestyle.

What autoimmune diseases and allergies have in common is that they are triggered by an overactive immune system. One popular theory as to why this occurs is known as the Hygiene Hypothesis. Crudely, the problem is that we are too clean.

The argument goes that because antibiotics and wet wipes have made our environment increasingly sterile, our immune system has responded by becoming overly sensitive. So, rather than just fighting genuine threats, like cholera, it treats things like pollen or gluten, which aren't threats, as if they were. We've cleaned up our world to such an extent that our immune systems, with nothing better to do, have gone on the rampage.

According to the Hygiene Hypothesis, your immune system is a bit like a grumpy teenager, bored out of his skull, who decides to start smashing the house up. The solution? Give him something to do, enemies to fight.

Some people have taken this to mean we should be deliberately exposing our kids to infectious diseases when they are young, while others use this theory to attack vaccination, arguing that it 'overwhelms' the immune system.

The trouble with the Hygiene Hypothesis is that it is probably wrong and doesn't fit the facts. In recent years it has been replaced by a subtler and more plausible version which focuses on the problem of the Old Friends we met earlier.

According to the Old Friends Hypothesis, the real problem is not that your immune system is bored, but that it is ignorant. Like us, our immune system comes into the world with an awful lot to learn. One of the things it needs to quickly learn is what is dangerous and needs to be fought, and what is OK and should be left alone. You really don't want your immune system launching an all-out attack on the microbes in your gut, since many of them are essential for your health.

In the past the immune system would have been taught these essential lessons by Old Friends, the gut microbes that evolved with us over millions of years.

Sadly, as we have seen, because of the overuse of antibiotics and a diet heavily based on highly processed foods, many of those Old Friends are no longer with us or only present in small numbers.

So the real problem is not that we are exposed to fewer infectious diseases (which are dangerous and harmful) but that we have lost contact with the microbes that evolved with us and without which our immune system just won't function properly. We have lost or accidentally killed off huge numbers of important Old Friends, and that is what is creating the modern problems of allergy and autoimmunity.

The next chapter of this book is all about how you can cultivate and re-establish contact with some of these Old Friends. But before that, I would like to introduce you to a gut parasite that has also evolved with us over millions of years and that has, in the past, played an important role in controlling and educating our immune system. It's not really an Old Friend but it can, in the right circumstances, be a valuable ally. I'm talking about parasitic worms.

Horrible hookworm

The hookworm is an unpleasant bloodsucking parasite that lives in the small intestine. It infects half a billion people worldwide and is a major health hazard in the tropics and subtropics.

Its life cycle is both fascinating and repulsive. It lives in human guts where it meets and mates with other hookworms. The female then lays around 30,000 eggs a day, which exit the gut in the faeces. Protected and fed by all this warm poo, the eggs hatch into larvae and wait patiently for a new host to come along.

If you happen to be strolling by, bare-footed, and step on ground that has been infected by the larvae, they will swiftly burrow in through your skin and travel up to your lungs. This will make you do a lot of coughing, which brings them up through your trachea and into your oesophagus. You swallow them, they somehow survive the warm acid bath of your stomach and make it to your small intestine. Here they latch on and begin to drink your blood.

Knowing all this, you might wonder why anyone would want to deliberately infect themselves with hookworm. The answer is, if you are desperate enough you will probably try anything.

I met Daniel Hayman in an Indian restaurant, where we both ordered a hot curry. This was something Daniel could never have contemplated doing a few years ago because he has Crohn's, a form of inflammatory bowel disease (IBD). IBD affects about 1.3m people in the US and, as with most autoimmune diseases, rates are rising. It happens when your immune system, possibly triggered by an infection, starts to attack the lining of your gut wall.

Unlike coeliac disease, IBD cannot be treated by the simple removal of an irritant. Like a hooligan who has decided to target your house, it keeps coming back. You may have periods of remission, but it is normally a lifelong condition.

Symptoms of IBD include pain, recurring or bloody diarrhoea, extreme tiredness, and weight loss. The first treatment you will be offered is usually steroids to calm things down. If that doesn't help, you may be given medication to suppress your immune system or to reduce inflammation. Some people end up having the inflamed bits of their bowel removed.

Daniel tried to control his disease with conventional medicine, but he continued to feel terrible, so he turned to something very unconventional – hookworm.

Daniel went on the internet and ordered live hookworm larvae from someone who had been cultivating them in his own body. When the larvae arrived Daniel placed them on his skin, where they burrowed in, eventually finding their way to his guts. The improvements were not immediate. In fact, it took almost a year before he felt fully well.

'Were you spooked by the idea of having a parasite inside you?' I asked him.

'Not at all,' was his surprisingly calm reply, 'I believe that human beings have evolved to have plenty of parasites inside them. It's part of life. You should try it.'

And then, just to prove how far he had come, he dug into one of the hottest curries on the menu. He is, he assured me, no longer afraid of food. 'I'm finally free. Thanks to the worms I can eat pretty much anything I want.'

So what's going on? I met up with Helena Helmby, from the London School of Hygiene and Tropical Medicine, who studies how worms interact with the immune system.

'We have a sophisticated immune system,' she told me, 'that's constantly on watch. But these large parasites have evolved with us for millions of years and they have developed ways to deal with it. One of the ways they do this is by dampening down our immune responses – and they do this by secreting compounds that manipulate our immune system. That enables the worms to survive, but it also has benefits for us. The worms are allowed to stay because any serious attempt to kill them would be far too dangerous to the host. So there's an uneasy truce between the worm and us.'

Helena agrees with the idea that one of the reasons we have seen this huge rise in allergies and diseases like IBD is that we don't harbour enough Old Friends. What she doesn't approve of is people deliberately self-infecting, as Daniel has done.

'Scientists have worked for over 100 years to eradicate these diseases and now people want to start reintroducing them? These worms live in the gut and suck blood. They move around. They shed eggs and larvae. If you are infected by too many of them you will become anaemic. Yes, we should be finding out how they work, but no, we shouldn't be ordering them up on the internet.'

Most scientists would agree with Helena that hookworm simply have too many downsides to be used as medication. Instead, they have been testing more benign parasites, like whipworm. Despite the yuck factor, there have been some successful clinical trials using these worms to treat IBD. In one study[16] they recruited 90 people and infected half with whipworm, and gave the others a placebo. None of the volunteers knew which group they were in. Result? Well, 43 per cent of patients who got the worms felt better afterwards, compared to 17 per cent in the placebo group. Not bad, but there's clearly some way to go.

Oddly enough, soon after I met Daniel Hayman I had the opportunity to deliberately infect myself with parasitic worms, in my case rather large ones. I had volunteered to swallow some tapeworm cysts as part of a research project done with scientists from Salford University.

The cysts I swallowed came from a beef tapeworm, *Taenia saginata*, which we got from an abattoir in Kenya. As parasites go, beef tapeworm is relatively harmless and only infects humans and cattle. Pig tapeworm, on the other hand, can be extremely dangerous, as the larvae sometimes migrate to the brain. That is why you should never eat raw or undercooked pork in some of the poorer parts of the world.

There is a widespread belief that tapeworm help you lose weight and in Victorian times it was relatively common for women to swallow tapeworm eggs in the hope of shedding a few pounds. This was almost certainly a waste of time as tapeworm eggs are not infectious to humans. They first have to be eaten by a cow or a bull, where they form a cyst, and it is only if you eat raw, infected meat with cysts in it that you are going to acquire a worm.

But even if you were infected, would you lose weight? I swallowed three tapeworm cysts that we cut out of the tongue of a dead cow (I wanted to be sure I got infected), but despite hosting three worms for a couple of months I actually put on weight.

It could be that the tapeworms were encouraging me to eat more, or it could be that I was unconsciously compensating for them being there. Either way, they don't seem to be a great weight loss aid, but nor did they create any noticeable problems for me.

In fact, they caused so little trouble that the first time I was absolutely certain that they were really down there was when, two months after swallowing the cysts, I swallowed a pill camera and saw them latched onto the walls of my small intestine, looking very much at home.

Before I started this particular self-experiment I had agreed with my wife that I would kill the tapeworms before they were fully mature. Once they are fully grown they start to shed segments, which crawl around looking for a new host. Understandably my wife was not keen to share our bed with any tapeworm segments which might have come out of me during the night.

One tapeworm expert I know, who deliberately infected himself as part of a research project, had the unpleasant experience of feeling a segment crawl down his trouser leg while driving along the motorway.

Anyway, once I had seen and filmed the young hookworms in my gut, and completed a round of tests, I took some pills and that was the end of the experiment. Although I kept on looking, the worms never came out so I can only assume that, when they had died off, my body treated them just like food and digested them. An ironic end – parasites eaten by their host.

Summary

- Microbes influence our immune system. The 'good' ones, our Old Friends, are particularly important when it comes to teaching our immune system how to behave.
- A lack of Old Friends leads to an overactive immune system and an increased risk of allergies.
- As we'll see in a moment, bolstering our Old Friends can help us stay healthy.

5

The Clever Guts Diet

Now we've done a quick tour of the marvellous, murky world of the gut, it's time to have a look at what you can do to keep your gut garden in good shape. In this chapter I will show you how you can improve your biome by changing what you eat and drink. I will also look at what you should avoid.

You are what you feed your biome

These days, when I put something in my mouth, I consider what it will do not just to my body, but also to my biome. Every food decision you make ('Shall I have that slice of cake or that handful of almonds?') decides the fate of countless billions living in your colon. It is quite a responsibility.

Although the following recommendations are based on the latest science, you should take them as guidelines rather than gospel. If you have IBS then you may have to avoid some of the foods we suggest, at least until you have got on top of your condition. See Chapter 7, How to Reboot your Biome, on page 186.

I'm a fan of whole grains, like rye, but they are not going to do you much good if you have a gluten intolerance or coeliac disease. And even if you have no gut problems, the response to exactly the same foods varies from person to person. So do experiment with these foods and our recipes, and see which suits you.

Having said all that, there are some foods that are better for you than others. The Clever Guts Diet draws on foods and spices from around the world, including the Mediterranean, India and Asia. For a diverse gut it is helpful to eat from a range of cuisines.

I'll start with foods most commonly found in my favourite way of eating, the Mediterranean diet. I love Mediterranean food, mainly because it's so tasty, but also because numerous studies have shown that this is an incredibly healthy way to eat.

One of the most impressive dietary studies ever done, the Predimed study, showed that, compared to going on a classic low-fat diet, going on a Mediterranean diet will cut your risk of heart disease, type 2 diabetes, breast cancer and even stop your brain from shrinking.[17]

For this study, which was funded by the Spanish government, researchers randomly allocated 7400 overweight Spaniards to either a low-fat diet (lean meat, low-fat dairy, minimal use of oil), or a Mediterranean diet (oily nuts, oily fish, lots of olive oil, eggs, dark chocolate, red wine). They were then followed for many years. The results were clear.

Although they were not calorie restricted, those randomised to the Mediterranean diet put on less weight, particularly around the middle, than those on the low-fat diet. They were 30 per cent less likely to develop heart disease and half as likely to develop type 2 diabetes. A particularly important element of the diet seemed to be olive oil. In fact, women in this study who were asked to add an extra couple of teaspoons of extra-virgin olive oil a day cut their risk of developing breast cancer by nearly 68 per cent, compared to those who were dutifully eating low fat.

Foods that nourish the gut

Olive oil

As well as adding flavour and taste, olive oil is one of the healthiest fats you can eat because it's rich in a range of polyphenols and antioxidants, which are good at damping down inflammation, whether in the brain, the breast or the gut.

Olive oil is one of those good fats that fill you up, so you'll be less tempted to eat processed fat and sugar later in the day.

Should you buy ordinary or extra-virgin olive oil? For me it comes down to cost, convenience and personal taste. I cook with the ordinary (contrary to popular myth, olive oil is one of the most stable oils you can fry with) and splash the extra-virgin on my salads.

Be aware, however, that not all extra-virgin is extra virgin. Because it is expensive to produce extra-virgin olive oil is now one of the two most counterfeited foods in the world (Manuka honey is the other).

The counterfeiting has been going on for many years, but the scandal erupted in 2010 when researchers at the University of Davis in California tested a wide range of imported extra-virgin olive oils, mainly from Italy. More than 70 per cent of the oils failed their tests either because they were rancid or had been adulterated by other, cheaper oils.

According to Tom Mueller, a journalist and author of *Extra Virginity: The Sublime and Scandalous World of Olive Oil*, this is a widespread problem, particularly in Italy, where it involves the Mafia and other crime syndicates. Popular ways to adulterate extra-virgin olive oil include adding canola oil or much poorer-quality olive oils. The blended oil is then chemically deodorised and sold on as extra-virgin oil.

Unless you have access to a well-equipped laboratory, there is no reliable way of detecting this adulteration. There are lots of urban myths about testing olive oil, such as putting it in the fridge to see if it solidifies, or trying to light it and seeing how it burns. None of these home-based tests work. If you want to find out more about which brands you can trust, and which you can't, do visit my website cleverguts.com, as this is a field that is evolving all the time.

Oily fish

Like olive oil, oily fish are full of good fats which are anti-inflammatory. In the case of oily fish, the key ingredient is omega 3. Unfortunately, white fish, like cod, don't contain many omega 3 fatty acids and they don't provide anything like the same health benefits. That said, even non-oily fish are a good source of protein. Not long ago 'experts' were worried about oily fish because of their high fat content and you still see some government websites recommending the non-oily versions.

One of my favourite maverick self-experimenters, Professor Hugh Sinclair of Oxford University, helped put the record straight.

In the 1940s Hugh Sinclair travelled to northern Canada to study the people living in the Arctic. Once known as Eskimos, they are now officially called the Inuit.

During his stay Hugh became intrigued by the fact that, although the Inuit ate lots of fat and very few vegetables, they had relatively low rates of heart disease.

He did further research and concluded that omega 3 – an essential fatty acid found abundantly in oily fish – was protecting the Inuit from heart attacks. He wrote a number of articles about this, but was largely ignored. So in 1979 he decided to put himself on an Inuit diet, consisting of seal, oily fish, molluscs and crustaceans.

He suspected, based on animal studies, that eating lots of fish oils would, at the very least, increase his clotting time. As the name implies, 'clotting time' is a measure of how long it takes your blood to clot. This is important in heart disease because it is the forming of a blood clot, which breaks off and blocks one of your arteries, which ultimately leads to a heart attack or stroke.

A few years ago I decided to repeat his experiment myself, though I had to skip eating seal (I ordered some from Canada but it got impounded in customs). I also couldn't stomach the idea of fish with every meal so I confined myself to one portion of oily fish a day. Hugh Sinclair was on his diet for three months; I managed one month.

Like him, before I started on my fishy diet I did some tests, which included measuring my clotting time. Over the course of the month I repeated the tests a number of times.

By the end of the experiment my 'bad' cholesterol levels had halved and my clotting time had doubled. Hugh Sinclair, on his more extreme diet, saw his clotting time go up from three minutes to a frankly terrifying 50 minutes. I say 'terrifying' because if you have an injury and start to bleed it is useful to be able to clot, otherwise you risk bleeding to death.

To cut a long story short, it is now widely accepted that eating oily fish at least twice a week reduces not only the risk of heart disease but also anxiety, depression and inflammatory conditions such as arthritis. But what effect does it have on your biome?

Well, following on in the footsteps of the great Hugh Sinclair, a human volunteer recently offered himself up as a guinea pig for trial by fish.[18]

The subject, who is just described in the study as 'a 45-year-old Caucasian male', had a reasonable diet before he started, which included red meat and vegetables. For a couple of weeks he changed to a fish-protein-only diet, which added up to just over 600mg of omega 3 a day.

During the course of his new diet, the researchers, naturally enough, took lots of stool samples. They saw what they described as 'a remarkable increase' in the levels of a number of bacteria associated with butyrate production. As I've pointed out before, butyrate plays a key role in maintaining gut health. It is the major source of energy for your colonic mucosa (the mucous lining that protects your colon) and it also helps reduce inflammation.

He then reverted to his normal diet. Unfortunately, after a couple of fish-free weeks, his poo samples returned to where they had been before. It seems that if you want to keep the benefits you have to keep eating the fish.

Good examples of oily fish include salmon, tuna, trout, sardines, mackerel and herring. Some people are anxious about the mercury found in fish, though I think the benefits clearly outweigh the risk. As a precaution, pregnant women are recommended to eat no more than two portions a week.

You can get more details about levels of mercury found in different fish from the US government's Food and Drug Administration website.[19]

Below are some average mercury levels, measured in parts per million. As you can see the fish to avoid are swordfish and shark (sharks are, anyway, horribly over-fished):

Swordfish: 0.995 ppm
Shark: 0.979 ppm
Canned tuna: 0.128 ppm
Cod: 0.111 ppm
Trout: 0.071 ppm
Mackerel: 0.050 ppm
Salmon: 0.022 ppm
Anchovies: 0.017 ppm
Sardines: 0.013 ppm
Shrimp: 0.001 ppm

You can get omega 3 fatty acids from other foods, such as grass-fed beef, walnuts, flaxseeds and hemp oil, but they come in a less biologically effective form. Lots of manufactured foods are now being fortified with omega 3, but because the omega 3 they use is extracted from algae it has a fishy aroma. Since potential customers would, understandably, be put off by things like fishy-smelling margarine, manufacturers mask the smell, chemically. The resulting food is unlikely to be anything like as healthy as you would hope.

As well as fish, an excellent natural source of omega 3 is seaweed (see page 126).

Fish oil capsules

What happens if you don't like oily fish? Well, there are always fish oil capsules. These are hugely popular but not hugely effective, at least not in humans. Most experts I've talked to are sceptical, and that's because the science doesn't support the claimed health benefits. A large meta-analysis of 89 studies published in 2006 showed no benefit from consuming fish oil capsules, and a more recent study from 2012 published in the *Journal of the American Medical Association*, covering 20 studies and 68,680 patients, also found that taking fish oil capsules made no difference to your chance of having a heart attack or stroke.[20]

Part of the problem may be the poor quality of some fish oil capsules. A research team from New Zealand examined fish oil capsules from all over the world and found that most contained fish oils that were oxidised – i.e. rancid. This was despite them being well within the sell-by date. 'Fish oil capsules' sounds wholesome, but it can take three years to get oils from the fish to those plastic bottles you see on pharmacy and supermarket shelves. Omega 3 is relatively unstable and in that time it may go off. Just as eating rancid fish is unlikely to do you much good, swallowing oxidised fish oil is unlikely to lead to a healthy biome.

The best way to tell if a fish oil is off is by taste and smell. If you get an unpleasant fishy burp after consuming a capsule you may have a dodgy pack. Do buy something with a trademark and keep the capsules in the fridge. Or you could just stop taking them altogether, as millions have.

As the *Washington Post* recently put it: 'Fish oil pills: A $1.2 billion industry built, so far, on empty promises.'

With all this bad publicity it's not surprising that the bottom is dropping out of the lucrative fish oil market, with sales falling internationally. If you really, really don't like fish, I would recommend cod liver oil. Cod liver oil is also an excellent source of vitamin D. I used to hate fish, but I have been converted. We have some fish recipes in this book that I hope will help change your mind.

Meat

So, oily fish is definitely on the menu, but what about meat, particularly red meat, like beef, lamb and pork? In recent years I've seen plenty of headlines along the lines of 'All red meat is risky' and 'If you want to live longer, hold the red meat'.

Really? On the plus side we know that red meat is a great source of protein. Beef, whole or minced, is also an excellent source of iron and vitamin B12, which are vital for producing healthy blood cells, and which many people are chronically deficient in. What's more, vitamin B12 is an essential nutrient for the developing brain, one of my favourite organs.

On the not-so-plus side, red meat is also rich in saturated fat. Beef has about nine times more saturated fat per gram than tofu. But is saturated fat as bad as is often claimed?

I recently met up with Dr Ronald Krauss, one of the world's most respected nutrition researchers. Dr Krauss is based at the Children's Hospital Oakland Research Institute in California. He has spent his professional life investigating the link between saturated fat and heart disease. He has very personal reasons for doing so.

'My father had a heart attack when I was very young,' he told me. 'This had such an impact on me that I decided when I was six years old that I was going to try and do something about heart disease.'

He trained as a doctor, then immersed himself in research. Because of his family history, he was concerned about his own health, so naturally he followed the dietary advice of the time, going on a low-fat diet and shunning cholesterol-rich eggs.

He also put his patients and research subjects on this diet. The trouble was, it didn't produce the results that they were expecting. 'We thought that everybody would get better on this diet, that their cholesterol profile would improve. In many cases they got worse.'

The problem is that when you go on a low-fat diet you have to eat something else. The 'something else' was often carbohydrates, like bread, pasta or potatoes.

Dr Krauss discovered that eating carbohydrates increases levels of a particularly damaging form of cholesterol known as small particle LDL (low-density lipoprotein).

Yet most doctors believed that saturated fat was the enemy. So with colleagues he decided to do a meta-analysis, pulling together as many high-quality studies as they could find. When, in 2010, they published their findings in the *American Journal of Clinical Nutrition*,[21] it was a bombshell. 'We found the heart attack risk was slightly higher but not significant. Stroke risk was actually reduced. So overall if you took it across the entire range of disease, heart disease and stroke, there was absolutely zero effect.'

Other more recent meta-analyses, such as one from Cambridge University, funded by the British Heart Foundation,[22] came to similar conclusions.

Dr Krauss doesn't see any harm in eating red meat a couple of days a week, although he prefers fish, topped up with vegetables and nuts. 'I stay away from potatoes as much as possible but my real weakness is bread. I have half a bagel most days, because I cannot resist them. Sugar almost never except for a little dessert with some chocolate. My main indulgence is a small piece after dinner because I consider chocolate an essential nutrient.'

Other American experts, like Dr Walter Willett, who is based at Harvard University's School of Public Health, are less enthusiastic about red meat.

'In our studies we've shown that those who consumed higher amounts of red meat had a higher risk of total mortality, cardiovascular mortality and cancer mortality,' he told me in the Harvard cafeteria, while I nonchalantly ate a steak. Dr Willett almost never eats meat.

According to Dr Willett, eating 85 grams a day – a small steak – increases your risk of premature death by 13 per cent.

This surprised me because a massive European study,[23] has come to exactly the opposite conclusion. In this study, which involved following nearly half a million people in 10 European countries for over 12 years, scientists found that eating moderate amounts of red meat had no effect on mortality.

If anything, it seemed to be protective. If you ate over 160g (5oz) of red meat a day, then, yes, you were at greater risk of heart disease and cancer. But those eating less were no worse off than non-meat eaters, and possibly a little healthier.

So how do you explain all these apparently conflicting studies? Well, one of the things that struck me was that the European study was done in Europe, where farmers are not allowed to routinely feed animals with antibiotics and growth promoters. In the US (where Dr Willett's studies are based) things are very different.

If you have ever visited one of those huge industrial cattle factories down in Texas, as I have, and watched the cattle being fed corn, antibiotics, growth promoters and a bit of green dye (to make the whole thing look more palatable), then you might not be surprised that Americans eating this sort of meat don't do too well.

I think what's important is not so much whether the meat you are eating is red or not, but what that meat itself has been reared on. If it has been brought up on antibiotics and growth promoters, it is unlikely to be doing you or your biome any good.

I continue to eat red meat, good quality and grass-fed, once or twice a week. But before carnivores go off rejoicing, I have to tell you that processed meat (like bacon, ham, sausages and salami) is quite a different story. More on that on page 156.

Three other foods that have been demonised but which are surprisingly good for your gut

Cocoa

Whoever invented milk chocolate was an evil genius. It is the most craved food on the planet. I have tried to give it up on many occasions and failed miserably. My kids know that if they leave chocolate anywhere visible around the house, I will probably eat it. I no longer buy milk chocolate because I know that once I start on a bar I simply will not stop. A few weeks ago I was driving home, alone, and stopped to buy some petrol. While I was there I 'accidentally' found myself buying a bar of milk chocolate. When I returned to the car I threw the bar of chocolate into the back, hoping to remove temptation. Instead, I thought about that chocolate for the next 10 miles and eventually had to pull over to eat it.

There are a number of reasons why milk chocolate is so seductive, including the flavour, smell and the way it melts in your mouth. And, of course, the marketing. But what it also has is the magic '50:50' ratio. Half of the calories in a bar of chocolate come from sugar, half from fat. You will find the same is true of a lot of the foods you crave, including doughnuts and ice cream.

Although we talk about sugar being 'addictive', few of us would eat a bowl of sugar by itself. Similarly, few of us would drink a glass of cream. But mix them together, add some flavourings, freeze, and you have ice cream, which is impossibly moreish.

Why is this 50:50 ratio so important? Well, perhaps because it was there in the first food you ever consumed. Breast milk has a 50:50 ratio of calories from sugar to calories from fat which you will find nowhere else in the unprocessed foods you eat. Melon is high in sugar, but not fat. Steak has lots of fat, but no sugar. It is one of those theories that would be hard to prove but certainly sounds plausible to me.

Whatever the reason, I am hopelessly hooked on chocolate. To overcome these cravings, my main strategy has been to switch to eating dark chocolate, which has less sugar and which I find far less addictive. I'm also pleased to say that dark chocolate is one of the things that the Mediterranean diet says is fine to eat, particularly if it has a high cocoa content. The problem with chocolate, healthwise, is not the cocoa, but the sugar. Cocoa itself is surprisingly good for you. It is broken down in your colon to produce nitric oxide, which expands your arteries and is good for your cardiovascular system. Cocoa is also an excellent source of flavonoids and polyphenols which, as we've seen, are good for your gut bacteria.

Eggs

I eat eggs most mornings. Not so long ago this was considered nutritional suicide. Like fat, cholesterol was demonised. If it was common sense that fat would clog up your arteries, then it was obvious that foods rich in cholesterol, like eggs and seafood, would do the same. We were urged to eat no more than a couple of eggs a week and lots of people started eating egg-white-only omelettes. Which was a terrible shame because almost all the nutrients are in the yolk.

Well, it turns out that like the fear of fat, the fear of eggs was also totally misplaced. A meta-analysis of 17 studies published in the *British Medical Journal* concluded that 'higher consumption of eggs is not associated with increased risk of coronary heart disease or stroke'.[24] Government guidelines around the world have now done a complete about-face, and we're all being encouraged to eat more eggs.

Whether scrambled, boiled, poached or in an omelette, eggs are a superb source of protein, rich in vitamins and minerals. There are around 90 calories in a boiled egg, half those in a small bowl of Frosties and a quarter of those in a croissant with butter and jam (400 calories, plus). Unlike the cereal or the croissant, the protein in the eggs will keep you feeling fuller for longer.

I can't find any studies which show that eating eggs will do great things to your biome, but if it stops you having sugary cereal for breakfast or scoffing that muffin mid-morning, then that is mission accomplished.

Wine

We all know that large amounts of alcohol are very bad for us, but what about more modest quantities? In the famous Predimed study, the volunteers who were randomised to the Mediterranean diet were allowed a glass or two of wine with their evening meal. Yet this flies in the face of much current medical advice, which states that any amount of alcohol is likely to do harm. So where does the truth lie?

A recent study in the US[25] supports the claim that modest amounts of alcohol can be healthy. In this study they followed more than 14,000 adults aged 45 and older for 24 years. What they found is that those who drank up to 12 units of alcohol per week (the equivalent of around six glasses of wine a week) had a lower risk of developing heart failure than those who never drank alcohol. But what about the biome?

Now, on the face of it, your microbiome should really hate alcohol. Pure alcohol is death to anything that is microscopic. That's why we use alcohol-infused swabs if we really want to clean our hands. We also know that alcohol can inflame the stomach, which is not a good thing. But alcoholic drinks like wine contain more than just alcohol. They also contain polyphenols.

As we've seen, polyphenols are chemicals found in tea, coffee, wine, fruit, vegetables and chocolate. When you eat these foods, around half of what you have consumed passes straight through the small intestine to the large intestine, where it has a powerful influence on the microbiome.

To see exactly what effect the polyphenols in red wine are having, Spanish researchers recruited 10 healthy middle-aged men and asked them to take part in an experiment.[26] After an alcohol-free week they were randomly allocated to either drinking a large glass of red wine (270ml), red wine with the alcohol removed, or gin (100ml) each day with their evening meal. After 20 days they switched regimes – each volunteer had to follow all three of them Throughout this experiment blood samples and poo samples were taken on a regular basis.

What the researchers found is that when the volunteers were drinking the red wine, and to a lesser extent when they drank de-alcoholised wine, there were significant drops in blood pressure, in C-reactive protein (CRP – a measure of inflammation) and in their triglyceride levels (the amount of fat circulating in the blood).

In addition, there was a marked change in their gut bacteria, with a particular increase in *Bacteroidetes*, the type of bacteria associated with slimness. They also noticed a significant increase in *Bifidobacteria*, which are associated with lowering cholesterol.

When it comes to booze, red wine seems to trump white wine, while spirits offer no obvious health benefits. As for beer, which is also known as 'liquid toast', I tend to avoid it because of its high carbohydrate content. If you are trying to lose weight or if you have sulphur-reducing bacteria in your gut, then beer is definitely a no-no.

Fruit and vegetables

Everyone knows that fruit and vegetables are good for us, but even those who manage to cram in their five a day or more, tend to stick to an incredibly limited range of them. Although most of us have realised that there is life beyond carrots, lettuces and tomatoes, we are not particularly adventurous in our eating habits. And that means our biome ends up being pretty limited as well. One simple way to increase your variety is to introduce more colour to your plate (remember Dave, with the IBS, and his beige diet? That, I'm sorry to say, is a very common cause of gut problems).

Apart from the fact that it makes your meal look more appealing, why should you go for colour? Well, the pigments that plants produce not only give them their colour but also keep them healthy, as they are made up of hundreds of different bioactive compounds or phytonutrients. Eating a wide variety of different coloured fruits and vegetables will give your gut bacteria something to chew on, as well as providing nutrients.

Phytonutrients

Phytonutrients, also known as phytochemicals, are concentrated in the skins of fruits and vegetables and are responsible for their colour, scent and flavour. They also help protect the plant from fungi and bacteria. The best-known phytonutrients are carotenoids, flavonoids and polyphenols. Carotenoids and flavonoids are found in yellow, orange, red, blue and purple fruits and vegetables (see the Phyto Salad recipe, page 224). Polyphenols are found in foods like cocoa, olives, dark chocolate, tea, coffee and red wine. Dried herbs are also spectacularly rich in them.[27]

Green

There are two types of green vegetable. 'Leafy greens', such as spinach, chard, lettuce and kale, are a good source of essential minerals, including magnesium, manganese and potassium. Spinach is also rich in folate and betaine, which help regulate homocysteine (high levels of which are associated with heart disease).

The other major class of green vegetables are the brassicas, which contain sulphur and organosulphur compounds. These include cabbage, cauliflower and broccoli. Sulphur is essential for the production of glutathione, an important antioxidant, as well as amino acids that are essential for building tissue and muscle.

But a word of warning: if you have IBD you may do well to avoid them. Although sulphur is great for healthy guts, it can aggravate inflamed ones.

There is a significant social downside to eating sulphur-rich foods, too. Even people with a normal gut sometimes find that eating members of the brassica family (particularly Brussel sprouts) can lead to awful smells. That's because a third of the population have high levels of sulphate-reducing bacteria in their guts, which turn the sulphur in their diet into hydrogen sulphide (which produces the smell of rotten eggs). If you evacuate a room after you've eaten a brassica or had a few pints of beer and a white bread sandwich (both of which also contain a lot of sulphur), now you know why.

Yellow, orange and red: carotenoids:
Fruit and vegetables that are orange, yellow or red are often rich in carotenoids. Carotenoids also give autumn leaves their vibrant colours. Foods rich in carotenoids include, not surprisingly, carrots. The type of carotenoid you find in carrots can be converted to retinol, an active form of vitamin A. The fact that vitamin A is important for healthy eyesight may explain why carrots are supposed to help you see in the dark. But you also find carotenoids in egg yolks, bananas, melons, tomatoes, peppers and squash. One particularly popular, carotene-rich fruit is the tomato. Thanks to its intense umami flavour, the tomato is the most popular fruit in the world. And no, it's not a vegetable.

Black, green, yellow: flavonoids

The word flavonoid comes from the Latin word flavus meaning yellow, and in a plant this versatile substance not only attracts insects for pollination but also helps protect it from harmful ultraviolet light. In humans, flavonoids may help combat allergies, inflammation and infection, as well as the consequences of yo-yo dieting. Despite the original meaning of the name, foods that have a high flavonoid content come in a range of colours and include parsley, black tea, red wine, apples, cocoa and peanuts.

Blue and purple

Blue and purple foods get their colouring from a type of flavonoid called anthocyanins. The more intense the colour the greater the concentration of anthocyanin. You'll find high levels of these particular flavonoids in blackberries, blueberries, purple carrots, red cabbage and the skins of aubergines.

Eating anthocyanin-rich blueberries may improve memory as we get older. They also strongly encourage the growth of 'good' bacteria such as *Bifidobacterium* and *Lactobacillus.*

White

Examples of phytonutrient-rich white foods include garlic, white onions, shallots and leeks. These are rich in alliums and allyl sulphur compounds. The most intensely studied of this particular group is garlic, which has been used down the centuries for medical reasons, as well as in cooking.

Although there is no compelling evidence that garlic will ward off vampires, it does appear to be quite good at killing 'bad' microbes. Traditionally it was eaten raw to treat coughs, colds and croup. A clinical trial, using a garlic infusion as a mouthwash, showed positive results, although the ensuing 'garlic breath' was a significant drawback.

Herbs and spices

Many dried herbs like thyme, oregano and basil are rich in polyphenols, while ginger and cumin have anti-inflammatory effects. But the spice with the greatest reputation for being gut-friendly is turmeric.

Turmeric

Turmeric is a yellowy-orange spice that is very popular in South Asian cooking. It gives a distinctive colour to dishes like chicken tikka masala and curried cauliflower soup. Multiple health claims have been made about turmeric. But do they stand up?

There are at least 200 different compounds in turmeric, but the one that's of particular interest to scientists is called curcumin. A recent review article looking at the evidence[28] concluded that turmeric has powerful antioxidant and anti-inflammatory properties. It is good at inhibiting the growth of 'bad' bacteria, parasites and pathogenic fungi, and it directly protects the wall of the intestine.

Most human studies have been done with high doses of turmeric or curcumin taken in capsule form. But how well does it work when taken in food? Recently, I was involved in an experiment in which we randomly allocated a group of nearly 100 volunteers to either a teaspoon of turmeric every day, a placebo, or a supplement. Those who got the turmeric cooked with it, added it to soup, sprinkled it on yoghurt or stirred it into warm milk. Not everyone was crazy about the flavour. 'Strong and lingering' was how some people described it.

Nonetheless, they stuck with it and after just six weeks we saw some significant changes in the activity of a number of genes in the turmeric-consuming group.

Professor Martin Widschwendter from University College, London, who ran the experiment, was surprised and delighted by the results. 'It was really exciting, to be honest,' he told me. 'We found one particular gene which showed the biggest difference. And what's interesting is that we know this particular gene is involved in three specific diseases: depression, asthma and eczema, and cancer. This is a really striking finding.'

A lot more research needs to be done, but I think turmeric is a spice that's well worth using. We have included it in a number of our recipes, including a delicious latte (see page 205). You will find that adding pepper increases its bio-availability – the amount of the active compound your gut absorbs. I like a sprinkle of turmeric with a touch of chilli and pepper in an omelette in the morning.

Fibre

Eating more fruit and vegetables also means you are going to be getting more fibre, which is normally a good thing as most of us don't eat anything like enough. In a typical Western diet, we consume less than half the recommended amount, which is at least 25g a day.

For most people, boosting fibre intake is likely to be good for the guts, the heart and the waistline. But if you have a diseased gut, adding more fibre to your diet may actually make you feel worse.

Prebiotics – treats for the Old Friends

Although fibre in almost any form is helpful, there are some types that are particularly important when it comes to keeping your gut bacteria in good shape. These are known as 'prebiotics'. A prebiotic is a special type of plant fibre which your body can't digest but which encourages the growth of 'good' bacteria in your gut. The good gut bacteria turn that prebiotic fibre into chemicals like butyrate. As I mentioned before, butyrate has been shown to have powerful anti-inflammatory effects insidc the bowel. Prebiotics act a bit like a fertiliser, boosting the growth of 'good' bacteria.

Inulin

This is one of the best known of the prebiotics. It is found in lots of different plants, but in high concentrations in only a few. As well as promoting a healthier biome and reducing constipation, eating food with lots of inulin in it can improve you bone health, as it enhances calcium absorption, and reduce your risk of heart disease, as it lowers blood triglyceride levels. Foods rich in inulin include:

1. Onions, leeks and garlic

Yes, it's them again. They are packed with inulin. Fortunately, I love this particular family of plants and cook with them on a regular basis. Our recipes are full of them. The Spanish make a wonderful tomato base called *sofrito*, which consists of garlic, onion, paprika, and tomatoes all cooked together in olive oil. It's delicious with chicken, fish or prawns. A treat for your family and your biome. We have a tomato sauce recipe in the back (see page 254).

2. Chicory

Chicory is normally eaten in salad. It has pale, slightly bitter leaves tinged with green or purple. This group also includes radicchio with its variegated red and white leaves. It is a member of the sunflower and daisy family, a close relative of lettuce and dandelions. In some countries chicory leaves are still crushed and used as a poultice to treat skin inflammations. The root is particularly rich in inulin – it accounts for nearly half its fibre. You are most likely to encounter chicory root as a caffeine-free coffee substitute.

3. Dandelion greens

These are good in salads and a great source of inulin. To make the leaves more palatable, you can blanch them to remove the bitterness or sauté them like spinach. They are less bitter before the flower head appears. Adding a handful to a salad gives it extra flavour and crunch. You don't have to go far to find them as they grow as a weed. You can also buy them from health food shops. Dandelions are used in traditional herbal medicine to treat infections and liver problems (you can make a bitter enzyme-inducing salad from a few simple ingredients, as listed on page 212).

4. Jerusalem artichoke

Over 70 per cent of the Jerusalem artichoke's fibre comes from inulin, so as much as 1 to 2 per cent of its weight is inulin, making it one of the richest sources of this prebiotic. It is also known as 'sunchoke' in the USA or 'fartichoke' because the high levels of non-digestible carbohydrate almost inevitably leads to some flatulence. If you are dining on them, leave the window open. If you aren't a regular vegetable eater or you have IBS, Jerusalem artichokes are probably one to avoid. We include a delicious Jerusalem artichoke soup in the recipe section (see page 218). In addition to inulin, Jerusalem artichokes provide lots of vitamin B1, iron, potassium and a decent amount of copper and vitamin C.

5. Asparagus

As well as being almost 3 per cent inulin, asparagus is surprisingly rich in protein. I like to steam it for a few minutes, then flash-cook it over a griddle and eat it hot, with lemon juice and butter. The only downside of eating asparagus is that a few hours later you you may produce strange-smelling urine. Speaking from personal experience, it has a distinct hint of sulphur, which can take you by surprise if you are not expecting it. Not everyone produces a strong smell in their urine after eating asparagus and not everyone can detect it.

There is a statistic for almost everything – studies in the UK and the US have come up with different figures for the percentage of the populations who produce strange smells after eating asparagus. A British study from 1987[29] found that only 43 per cent of British males produce an obvious smell after eating asparagus, and it runs in families. A US study[30] found that 79 per cent of Americans produced a detectable odour.

6. Bananas

Bananas contain modest amounts of inulin, less than 1 per cent of their weight, but they are rich in a number of minerals and vitamins, though they also include a fair amount of sugar too. It's particularly worth including green or unripe bananas in your diet because they are a good source of resistant starch, which is another very important prebiotic.

Resistant starch (RS)

Resistant starch is a type of starch that, as its name implies, resists digestion in your stomach and small intestine and reaches your colon largely intact. You are unlikely to get big blood sugar spikes after eating RS, and you won't hold onto many calories from it either. Once it reaches the large intestine, it feeds the 'good' bacteria which digest it and release butyrate. As we've already seen, butyrate reduces inflammation and strengthens the gut wall.

You'll find lots of resistant starch in grains, seeds and legumes. You also get it in unripe bananas and green peas.

But one of the more surprising places you will find some of it is in pasta or rice that has been cooked and cooled.

I first became aware of this curious finding when researchers from a programme I worked on asked Dr Denise Robertson, from the University of Surrey, to run a slightly odd experiment. Volunteers were asked to eat pasta with a tomato sauce either hot, cold or reheated. Their blood sugar levels were measured after each meal and we discovered, as we'd expected, that eating cold pasta led to a smaller spike in blood glucose and insulin than eating freshly boiled pasta. That's because cooking and cooling pasta changes the structure of the starch in the pasta, making it more resistant to digestion.

But then we found something that we didn't expect – cooking, cooling and reheating the pasta had an even more dramatic effect. Or, to be precise, eating reheated pasta resulted in an even smaller effect on blood sugar levels than eating cooled pasta. In our volunteers it reduced the rise in blood sugar levels by 50 per cent. This strongly suggests that reheating the pasta made it into an even more resistant starch. As far as I know it's something that had never been measured before.

Since we ran that experiment, a group in Sri Lanka have done something similar with rice. They found that if they boiled rice with a bit of coconut oil, cooled it down, then reheated it in a microwave, they could increase the levels of resistant starch in it roughly 15-fold. They think that eating rice this way probably halves the amount of calories your gut will absorb from it.

At the very least it is a good excuse to use up any left-over pasta or rice you may have in the fridge. And by the way, the claim that reheating rice is dangerous is an almost total myth. Unless it has been sitting out in a warm place for many hours it should be absolutely fine.

For a list of foods with the most resistant starch, see page 268.

Other good sources of fibre

Barley

Barley is an ancient and very tasty grain, with a slightly nutty flavour. It is good in soups and stews, really giving them a bite. Barley scores well as a prebiotic because it contains lots of beta-glucan. Like inulin, beta-glucan promotes the growth of friendly bacteria in your gut. Beta-glucan binds LDL or 'bad' cholesterol in your gut, preventing it from being absorbed through the wall lining, thereby lowering levels in the blood.

Oats

Like barley, whole oats contain large amounts of beta-glucan fibre, as well as some resistant starch. I did a small experiment, in which I ate oats every morning for a month and discovered that it cut my bad cholesterol by about 10 per cent. Using whole oats, the sort of porridge that takes longer to cook, will also slow the release of the starchy carbohydrates and sugars, and the fibre is more likely to get to feed the good guys in your biome, whereas more processed, quick-cook porridge is less filling and will produce a greater sugar spike. (See our granola recipe on page 198)

Flaxseeds

This is another very healthy prebiotic, a seed with a slightly nutty flavour. You can scatter flaxseeds on your porridge in the morning, or try adding them toasted to salads for a delicious crunch. Flaxseeds are a reasonable source of omega 3 though, unlike fish, little of it is in an active form that your body can make use of. They are, however, rich in insoluble fibre, like cellulose. As well as feeding your microbiome, the fibre in flaxseeds should ensure regular bowel movements. (See our recipe for flaxseed bread on page 210).

Apples

I love apples and they are one of the healthiest fruits you can eat, as long as you consume them whole and not as a juice. They are another good prebiotic, low in sugar and high in pectin. Eating apples increases the production of our old friend butyrate, the short-chain fatty acid that feeds our 'good' bacteria. They are also high in polyphenol antioxidants. I love stewed apples, with yoghurt, or diced and cooked in the oven with a scattering of cinnamon (see page 260).

Seaweed

Slippery, slimy, and with a very distinctive ocean flavour, seaweed is not something I'd had much experience of before researching this book. It is a popular food in many parts of the world, particularly Japan, China, Korea and Taiwan. Seaweed comes in a wide range of colours including green, brown and red. Modern land plants are all derived from the green species of seaweed, which gives you some idea of just how diverse these sea plants are.

The seaweed species you are most likely to have encountered on your plate is nori, which is used to make sushi. Other common types include dulse, which I was told tastes like bacon when fried – it doesn't; kelp, which is used as a gluten-free alternative to noodles; and spirulina.

Not all seaweeds are edible, and some are poisonous, so unless you know what you are doing it is not a great idea to pluck it from the sea shore and wolf it down.

But the seaweeds that are edible are excellent prebiotics, packed with vitamins and minerals, as well as fibre and omega 3 fatty acids. With all this going for them it's not surprising that they have been hailed as yet another superfood. But where's the evidence?

Well, there have been lots of animal studies that have suggested benefits. Researchers from University College, Dublin, for example, have shown that feeding seaweed extract to sows produces long-lasting effects on their offspring. Their piglets have less of the nasty bacteria, such as *E. coli*, in the gut; they need fewer antibiotics during their lifetime and put on weight far faster than piglets fed on a standard diet. But what about human studies?

Dr Pia Winberg, who works at the University of Wollongong in Australia, describes herself has a seaweed farmer. She is incredibly enthusiastic about the potential of seaweed, pointing out its long history: 'Seaweed is one of the oldest "plants" used by humans and records reach back to Tasmanian aboriginal communities. People have been eating seaweed for thousands of years.'

She also says that you will find seaweed in toothpaste, paint, printer ink, ice cream and salad dressing. Scientists are even testing seaweed gel nasal spray products tested against bird flu. But it's the impact of seaweed on the microbiome that really interests her, something she has been assessing not just in the lab but at home.

'I've done lots of tests on my kids and everyone else I can lay my hands on. Blood, wee and poo are always exciting for kids.'

She has just completed one of the first human trials on

the effects of seaweed on the gut, which she plans to publish in 2017. For her study she recruited 64 people who were either overweight or obese. They were then randomly allocated to consume one of three capsules every day for six weeks. One of the capsules contained 2g of a special seaweed fibre, the second capsule contained 4g, and the final capsule was a placebo.

They volunteers went away and swallowed a capsule every day for six weeks, with neither they nor the scientists knowing who was consuming seaweed and who was getting the placebo. Blood and stool samples were taken at the beginning of the study and then again at the end. So what did they discover?

'What we found,' Pia told me, 'was a significant reduction in inflammation and an increase in insulin sensitivity. What we also found, when we looked at their stool, was that a group of about 15 different bacteria had increased in the people who were on the active seaweed supplements.'

These bacteria were mainly from groups which produce short-chain fatty acids like butyrate, which in turn encourage the growth of the mucous lining.

But what Pia found particularly interesting were some of the other benefits.

'Thanks to the increased fibre, everyone taking the seaweed supplements reported becoming more regular. I hadn't appreciated before the study how many people are constipated. At the end of the trial, when they found out what they had been consuming, people wanted to keep on taking the capsules, just for that reason.'

The volunteers taking the seaweed capsules also reported feeling less desire for sugar and fast carbs, which Pia thinks may be because the seaweed had changed their gut flora. This in turn could explain the improvements she saw in their insulin levels.

Did the seaweed make any of them gassy? I asked.

'We didn't find that in this study,' Pia replied, 'in fact a couple of participants with IBS reported that for the first time they could eat cabbage and other brassica-type vegetables. That is anecdotal but interesting.'

'The most surprising and stunning result,' she went on, 'was with a woman who had severe psoriasis. It was so bad on her hands that the skin just constantly flaked off. She had been in bandages for two years and lost her job. Four weeks into the study she took the bandages off and her skin looked fine. At the end of the study she stopped taking the capsules. Three weeks later her hands started peeling again.

'When she was taking the capsule the inflammatory factors in her bloods went down, so is seems likely that the reduced inflammation was helping her skin. We've tested her five times now, putting her on and off the seaweed, and each time she is off her hands peel. When she goes back on the psoriasis goes.' Pia and her team and now extending their clinical studies, looking specifically at people with psoriatic skin disorders.

The main downside to seaweed is that it has a fishy taste. Nonetheless, we have included a number of recipes in this book that have been tested on some very picky consumers, including my children, and had the thumbs-up. If you want to find out more about Pia's work and how to buy capsules containing seaweed dietary fibre, visit our website. When Pia publishes her study, we will post a link to it.

Probiotics – how to introduce more Old Friends

Unlike prebiotics, which are like fertilisers for your biome, probiotics are the live bacteria or yeast that you parachute into your intestine, in the hope that they will take root and do you some good.

To have a chance of reaching your colon alive, probiotics first have to survive a dunking in the acid bath of your stomach, followed by a dousing in gastric juices and a lot of pounding along the way. Even if they reach the colon they still face a vast army of microbes which are already in residence and unlikely to welcome any newcomers.

There's a lot of hype around probiotics, with lots of different supplements being advertised on the internet and in health food shops. Only a few have decent science behind them, and I will come to those in a moment. Unless you have a particular gut condition you are trying to treat (like antibiotic-induced diarrhoea), I think the best way to top up your levels of 'good' bacteria is through food. And one of the best probiotic foods is homemade yoghurt.

Yoghurt

I've made yoghurt many times – it is so much nicer than the shop-bought stuff, as well being richer in bacteria like *Lactobaccillus* (see our recipe on page 206).

When I buy yoghurt it is usually live and full-fat. I add fruits to sweeten it, or scatter cinnamon, flaxseeds and nuts over it. The sugary low-fat versions are unlikely to have any health-promoting properties. The evidence that full-fat is better for you than low-fat (even when it doesn't have sugar in it) is compelling.

Dr Mario Kratz, a nutritional scientist who is based at the Fred Hutchinson Cancer Research Center in Seattle, has looked at an awful lot of studies comparing the benefits of low-fat versus high-fat (i.e. normal, unadulterated) dairy and concluded that 'none of the research suggests the low-fat-dairy is better'. In fact, there are plenty of studies that have found that eating full-fat dairy, particularly yoghurt, is likely to lead to less obesity and a lower risk of diseases such as type 2 diabetes.

Yoghurt first acquired its healthy aura thanks to Ilya Metchnikoff, a professor at the Pasteur Institute in Paris, who wondered why Bulgarians from the Rhodope Mountains live, so much longer than most other Europeans. He decided it was because of bacteria in the yoghurt. Dr Kellogg, who helped found the famous cereal company, was so impressed by this research that he began recommending it to his patients. Except that he suggested they take it as an enema since he thought that was the best way to get it to the large bowel where it would do the most good.

I'm not sure about yoghurt enemas, but taken the normal way, through the mouth, there's certainly good evidence that yoghurt, with live or active cultures, can help treat conditions like antibiotic-associated diarrhoea and IBS, and help reduce constipation. If you want to find out which particular brands have been tested, go to our website.

Another good thing about yoghurt is that the fermenting process breaks down lactose so it contains much less than milk. Even if you are lactose-intolerant you should be able to eat yoghurt. In fact, eating it can alleviate symptoms.

Cheese

I love cheese but I gave it up for many years because it contains lots of saturated fat and this, I assumed, was likely to lead to an early grave. (I tried eating low-fat cheeses but they tasted terrible.) Then I looked into the science and realised there is no evidence that eating low-fat cheese has any particular health benefits over eating the full-fat version. So I returned to eating proper cheese and have enjoyed every mouthful ever since.

As anyone who has been to France knows, the French eat huge amounts, something like 20kg per person per year – more than any other country. Yet they have relatively low rates of heart disease and obesity. This is known as the

'French paradox'. It could be something to do with the fact that they drink lots of wine or maybe there is something special about their lifestyle, but gut bacteria may also play a part.

In a recent study, scientists recruited 15 healthy young men and asked them to eat lots of cheese, full-fat milk or a control diet for two weeks. During the weeks when the young men drank milk or ate cheese, there were higher levels of butyrate and propionate in their stools. Both are compounds produced by the biome that have been shown to be good for the gut.[31]

Not all cheese contains live bacteria, and processed cheese contains virtually none. It is important to look for live and active cultures on the food labels. You will find significant numbers of 'good' bacteria in Gouda, mozzarella, Cheddar and cottage cheese, as well as in blue cheese such as roquefort, though you will need to read the labels to be sure.

One of my other favourite cheeses is feta. It is often made with sheep's or goat's milk, and has a lovely creamy, tangy flavour. Feta is rich in a type of bacteria called *Lactobacillus plantarum*, which produces anti-inflammatory compounds. Because it is a fresh cheese made from whole milk it also contains quite a lot of lactose. If you are lactose-intolerant, this may be one to skip.

I love feta in Greek salad (see page 221). It is also delicious grilled or in an omelette with spinach and tomatoes.

Fermented foods

Sauerkraut, kefir and kimchi

Wine, cheese, yoghurt and chocolate are all fermented foods. That simply means they've been created by microbes converting sugars, such as glucose, into other compounds. The unique flavours and textures you get with different fermented foods are due to the different species of bacteria and yeast involved. I've noticed that sourdough bread made in San Francisco has a completely different taste from sourdough bread made in London and that has to be because the microbes in London are very different from those in San Francisco.

Fermentation has become increasingly popular, because it not only adds flavour and texture, but also adds lots of 'good' bacteria to your diet. Manufacturers have, of course, jumped on the bandwagon and started selling all sorts of fermented foods. The problem is that many of them have been pasteurised, so there's nothing living in them. I was recently involved in a study in which we sent off a number of shop-bought fermented foods, such as sauerkraut and kimchi, to a university to be tested. Sixty per cent of the samples we sent grew nothing. You need to buy the products fresh and preferably with some guarantee on the label that they really do contain living bacteria. Or, of course, you could make them yourself.

I'm not familiar with the whole fermenting scene, so I met up with a group of people who are. They're based near London and call themselves The Fermentarium. They are on a mission to spread the word about the joys of making your own fermented food.

When I met Gaba, who is originally from Poland, she was busy making sauerkraut. This is traditionally made from finely shredded cabbage covered in salty water and left to ferment. It is popular in Europe, where it is eaten with sausages or as a side dish.

Gaba learnt how to make it from her mother. 'My mum, 82, still makes sauerkraut,' she told me. 'When we were children, living in Poland, she would make a huge vat of sauerkraut in November and put it outside on the balcony, where it was well below freezing. When it came to mealtimes she would take a bowel, crack the ice, scoop up some sauerkraut, carry it inside and warm it up. And that vat of sauerkraut would see us through until the spring.'

Gaba said that while most people, particularly at first, tend to stick to fermenting milk or vegetables, you can ferment almost anything, including pork fat and raw fish.

At this point I had a flashback of one of the most revolting foods I have ever eaten, and that was fermented shark. I was in Iceland and because it is something of a tradition there I thought I would give it a go. Big mistake. I got one mouthful down before I started to gag. This is strictly for hardcore fermenters, though at least the fermented shark they eat these days is produced more hygienically than it used to be. In the good old days, I was told, you would take your freshly caught shark, put it in a bucket, urinate over it, add some salty water and then leave it somewhere in the basement to ferment until it was time for the party to begin. Which could be months.

Gaba assured me she never goes that far when she is fermenting fish. 'We don't do it to the point that the Icelanders and Norwegians do, not until it is rotten. We just ferment for a couple of days.'

Gaba adores the smells and flavours that you get only from fermented food. 'Why,' she wondered, 'do the British have no real culture of producing fermented foods? You make cider, apple cider vinegar, bread and cheese, but as a nation you seem to be very reluctant to eat, let alone make, the more exotic and vinegary rich foods that are found widely in Eastern Europe and further east.'

She thinks one of the main reasons why Eastern Europeans are keen on fermented food is that they have to endure much longer and colder winters. In earlier times it was almost impossible to buy vegetables for much of the year. That's why peasants, before the days of refrigeration, found so many different ways to pickle and preserve food.

The homemade sauerkraut Gaba had brought along for me to taste was delicious and nothing like the vinegary stuff from supermarkets that I've tried before. Freshly made sauerkraut is packed full of live bacteria and surprisingly easy to make. There's a recipe for sauerkraut on page 257, as well as one for kefir (a fermented milk drink, which offers one of the most probiotic-rich concoctions available) on page 208.

Kimchi, a spicy Korean side dish, made from vegetables, is another popular fermented food. I read somewhere that kimchi is going to be this year's Hot Ingredient, being added to everything from tacos to pizza, even to a grilled-cheese sandwich. It's versatile and packs a punch. Are your tastebuds and microbiome ready for it? Try it and see.

One of the reasons why fermented foods are so good for the gut is that, gram for gram, they contain a huge number of different microbes. The microbes in fermented foods are also far more likely than most other bacteria to make it safely down into your colon because they are extremely resistant to acid, having been reared in an acidic environment. You might worry about the impact on your gut of eating foods which are acidic, like sauerkraut and kimchi, but all the evidence shows that they are good for you. Do, however, start slowly if you haven't eaten these sorts of food before.

Apple cider vinegar

You may not have come across kimchi or kefir, but you will probably have tried apple cider vinegar. This is another fermented food that is now hugely popular, and we have included a number of recipes that contain it.

The actress Scarlett Johansson apparently washes her face with a diluted version of the stuff, while the Arctic explorer Sir Ranulph Fiennes also swears by it, claiming he beat his arthritis thanks to a vinegary drink (four parts apple cider vinegar to one part raw honey).

More heavyweight endorsement comes from Hippocrates of Kos, often described as the 'Father of Western Medicine'. Hippocrates was a great believer in the power of the body to heal itself. To help the body on its way he often prescribed vinegar and fasting. He described fasting as 'the physician within' and claimed that 'to eat when you are sick is to feed your sickness'. As for vinegar, he recommended its use for a wide range of things, from cleaning wounds to sorting out a persistent cough. But does it really have any remarkable health-giving properties?

I recently teamed up with Dr James Brown from Aston University to find out. The first thing we decided to test was the claim that drinking a couple of tablespoons of vinegar before a meal will help control your blood sugar levels and prevent blood sugar spikes.

We started by recruiting some healthy volunteers and asked them to eat a bagel. James measured their blood sugar levels before and after.

As expected, bagel consumption was followed by a large and rapid rise in their blood sugar levels. The next day we asked them to eat another bagel, but this time they had to knock back a shot of dilute apple cider vinegar just before doing so. This reduced the amount of sugar going in to our volunteers' blood by almost 50 per cent.

James was delighted with this finding. 'There have been studies which suggest this happens in people who are at high risk of diabetes,' he told me, 'but I'm not sure it's been shown in healthy volunteers before so I was pleasantly surprised.'

This probably happens because the acetic acid in the vinegar suppresses the activity of an enzyme called disaccharidase, which breaks down starches and complex sugars, enabling them to be more rapidly absorbed by the body. If

Homemade Yoghurt, p.206

Pumpkin Porridge p.197;
Breakfast Bread with Raspberry Chia Jam p.211; Chia Pot p.201

Classic Greek Salad p.221; Green Pea Soup p.221;
Walnut and Red Pepper Dip p.214; Chocolate Aubergine Brownies p.259

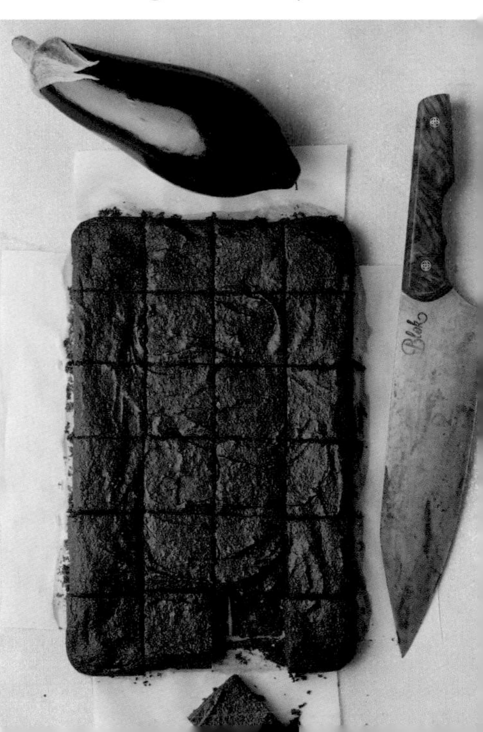

Phyto Burst Lunch Box p.226

The magic ingredients of Tanya's Gut-soothing Bone Broth, p.216

Chicken Goujons with Hummus p.244; Asian Coleslaw with Chicken p.235; Harissa and Beetroot Salmon p.245; Cod in Salsa Verde p.246

Baked Rainbow Ratatouille, p.254

Foods to enjoy on the Clever Guts Diet

the enzyme is blocked, far less sugar gets absorbed.

But would apple cider vinegar drinkers come out better in other measures, such as weight loss, lower cholesterol or reduced inflammation?

James divided volunteers into three groups. For eight weeks, the first group had to consume two tablespoons of apple cider vinegar just before lunch and dinner, either mixed with water or on a salad. The second group were asked to do the same with malt vinegar and the final group got vinegary-smelling water (placebo).

By the end of eight weeks none of them had lost any weight, though the apple cider drinkers were very positive about the experience. A woman with mild arthritis told me, 'I have had less aches and pains in my joints, especially after exercise.'

As part of the testing James had measured their levels of CRP, the marker for inflammation in the body. He didn't find major changes across the groups, although some of the apple cider vinegar drinkers did have falls in their CRP levels. What he did see, which surprised him, was quite a big change in cholesterol, but only in the apple cider vinegar drinkers.

In this group there was an average 10 per cent reduction in total cholesterol, with a particularly big reduction in triglycerides (a form of fat, carried in the blood). This was a striking finding because our volunteers were all healthy at the start, with normal cholesterol levels.

The fact that we only saw this effect with the apple cider vinegar, and not with the group drinking malt vinegar, suggests that the apples are playing a special role.

'Apple cider vinegar has lots of different, bioactive molecules which are found in apples,' James said, 'and at least two of those molecules have been shown to have really beneficial effects. So it's likely that there is a component of apples which is found in a concentrated form in apple cider vinegar which has given us this result.'

In this particular experiment we didn't look at the participants' poo, but other studies have suggested that apple cider vinegar can be good for the stomach as well as blood sugar and cholesterol levels. That said, it is not a cure-all and it is really acidic, so I wouldn't recommend splashing it on your face or glugging it down neat. Drink it diluted or use it sparingly in your food. I personally like it on salad, as a vinaigrette.

Clever Guts Apple Cider Vinaigrette

1 tbsp French mustard
1 garlic clove, crushed
2 tbsp apple cider vinegar
1 tsp balsamic vinegar
6 tbsp olive oil

Other uses for apple cider vinegar that I've heard of but not personally tested include:

- Wart removal – soak a cotton pad with vinegar and secure it over the wart. Leave it on overnight, and repeat every day for a week.

- Curing bad breath – add ½ tbsp to a cup of water, and gargle.
- Soothing a sore throat – mix 2 tbsp with 2 tbsp warm water and gargle.
- Shiny hair – add 4 tbsp to a cup of water and pour it over your hair after shampooing. Leave it on for a moment and then rinse.

Probiotic supplements

As well as in fresh foods, you can get probiotics in the form of capsules, pills or processed foods that contain live bacteria. There is a huge market for probiotics, many of which are over-hyped, like fish oil capsules. The response of most of the experts I've spoken to has been scepticism, particularly when it comes to so-called probiotic yoghurts like Yakult or Actimel. These are yoghurts with 'added bacteria'. The clever thing the manufacturers did was fund research into the potential health benefits of their products, on the back of which they were heavily promoted around the world.

But when scientists from the European Food Safety Authority looked at some of these claims they were not convinced. On Yakult's or Actimel's UK websites, you will see they mention that their products contain lots of *Lactobaccillus* but they are careful not to make specific health claims.

I personally avoid all such products, not least because they tend to contain extra sugar or sweeteners. But if I had a particular condition I wanted treated, such as antibiotic-induced diarrhoea, I might take a proper probiotic.

Graeme Close, Professor of Human Physiology at Liverpool John Moores University, certainly thinks that probiotics, targeted at specific problems, are worth considering. As well as doing his university work, he advises the England rugby team, Everton Football Club and the European Tour golfers on what to eat. He says that when his teams are all together in camp, training hard, taking a good probiotic may reduce the risk of them coming down with something nasty.

'What we find,' he said, 'in elite athletes is that after an intense exercise session you can have temporary damage to the gut, which creates a window of opportunity for pathogens, "bad" bacteria, to take hold. If you are training hard, day after day, you get these spikes of increased vulnerability.'

Which could explain why elite athletes are often ill before a big event. When world champion runner Paula Radcliffe went for Olympic gold in Athens, she was thwarted by a bout of severe diarrhoea during the race.

To minimise the risk of this happening, Graeme recommends taking a multi-strain probiotic if you are doing heavy training (for more details, go to our website). Graeme also recommends taking vitamin D supplements, particularly during the winter months, because low levels of vitamin D are bad for the biome.

'I don't recommend many supplements. I believe in food first,' Graeme assured me. 'In sports nutrition this has changed enormously. If you were in pro sport 10 years ago you could have been taking dozens of supplements.'

For the scientific evidence behind a wide range of probiotic supplements, I recommend you go to Medline Plus, a service provided by the American National Library of Medicine, which brings together reliable information from the National Institutes of Health as well as other health-related organisations: https://medlineplus.gov/druginfo/natural/790.html (and see page 270).

Foods to avoid

As well as things you should eat, there are also things you should cut down on and preferably avoid if you want a healthy, diverse biome. Top of my list are sugar and processed food.

Sugar

The high-sugar diet that many of us have been indulging in has had a catastrophic effect on our health. In the UK the average six-year-old eats their own weight in sugar every year, and attempts to curb this have been largely unsuccessful. If you still think that sugar is a bit of harmless fun then let me spoil your day by listing some of the downsides of this pure, white and deadly substance:

1. Sugar contains no essential ingredients. None. It's just calories. Other foods may have traces of vitamins, minerals, fibre… something. Sugar does not.

2. Sugar rots your teeth. I've had almost every tooth in my face filled and drilled because when I was a child I ate and drank too much sugar. The main reason why children under 10 are admitted to hospital is to have their severely rotten teeth extracted.

3. Eating or drinking sugar causes your blood sugar

levels to spike, then crash, which makes you hungry. People who eat a lot of sugary things tend to be fatter. This is particularly true of children. One study found that drinking a single sugar-sweetened drink per day increased the risk of becoming obese by around 60 per cent.[32]

4. Because of its effect on your weight and on your insulin levels, consuming lots of sugar puts you at increased risk of type 2 diabetes. This can lead to dementia and amputation. Being a diabetic also cuts 10 years off your life, despite medication.

5. Sugar is addictive. Not, oddly enough, on its own. I don't dive into the sugar bowl. It's when sugar is mixed with fat, in that ratio of one calorie of sugar to one calorie of fat, that we find it irresistible. There are some people who can take a sliver of cake and leave the rest, but there are plenty more who can't.

Food and drink manufacturers, who have spent years funding research to show that sugar is absolutely fine (a tactic reminiscent of the tobacco industry), are well aware that the tide is turning. So they have begun vigorous attempts to sell us low- or zero-calorie alternatives. But are these alternatives any healthier? The jury is out, but evidence is mounting that they are not.

The whole point of zero-calorie sweeteners is that they give you the taste of sugar without the calories. But your body and brain are not so easily fooled. A study on fruit flies, done by researchers from the University of Sydney, showed how artificial sweeteners lead to changes in parts of the brain which regulate appetite.[33]

They found that fruit flies consumed 30 per cent more calories after eating artificial sweeteners than after a food sweetened with normal sugar. So what's going on?

Researcher Greg Neely explained 'When sweetness versus energy is out of balance for a period of time, the brain recalibrates and increases total calories consumed.'

In other words, if your brain thinks you are about to get a sugar hit (because your taste buds tell it that sugar is on the way), but nothing arrives, then it will eventually decide you need to eat something else to compensate.

Fruit flies are, of course, not human, but these findings are in line with other studies that have shown that consuming artificial sweeteners, mainly in the form of fizzy drinks, is no better for you, and may even be worse, than taking your sugar straight.

That was certainly the conclusion of a big meta-analysis, looking at a whole range of recent studies.[34]

Artificial sweeteners

As well as affecting your brain, artificial sweeteners can change your gut bacteria.

A few years ago the two Erans from the Weizmann Institute, whom we met earlier, Dr Eran Elinav and Dr Eran

Segal, did experiments looking at the impact of consuming artificial sweetners on the biome.[35]

Mice were given water, sugary water or water laced with saccharin to drink. Unlike those that drank water or sugary water, the mice that got the saccharin soon developed glucose intolerance, a step on the way to diabetes.

Next, they transferred the biome from the mice that had consumed saccharin to 'germ-free', or sterile, mice. Within six days these mice had also developed glucose intolerance, suggesting that there was something in the transplanted biome that had been changed by the sweetener. A closer look at their poo confirmed that their guts now had more of the sort of bacteria that lead to obesity and diabetes and less of our old friend, *Akkermansia.*

Finally, they asked a group of human volunteers, who normally don't eat or drink artificially sweetened foods, to consume a high, but safe, dose for a week.

Within seven days many, but not all, of the volunteers began to develop signs of glucose intolerance. Why? Well it appears that some gut bacteria react to sweeteners by secreting chemicals that provoke an inflammatory response, which in turn encourages obesity and diabetes. If you have those sorts of bacteria, sweeteners should definitely be off the menu. How can you find out if you have them? At the moment you can't. I steer clear of artificial sweeteners and would recommend you do as well.

Dr Elinav, who is clearly not a fan of artificial sweeteners, thinks that there should be a 'reassessment of today's massive, unsupervised consumption of these substances'.

So what's the alternative?

Unfortunately, natural alternatives to sugar, like honey and agave syrup, will have a similar effect on your blood sugar levels as eating sugar. I'm not totally against them, but they are a wolf in sheep's clothing.

The trouble is, I still like the sweet stuff. So I was intrigued when someone suggested I try the extract of an African fruit, *Synsepalum dulcificum,* better known as the miracle berry. Miracle berries contain miraculin, a molecule that binds to receptors on your tongue, changing their shape. Miracle berries are unlike any other sweetener because they work not by making foods sweeter but by making them *taste* sweeter.

Fresh berries are hard to get your hands on, but you can buy the dehydrated pulp of the fresh berries in tablet form on the internet.

So what are they like? I put one on my tongue, waited for it to dissolve and then I was good to go. I had read enthusiastic claims that it made foods, such as oranges, taste as if they had been 'freshly plucked from the Garden of Eden'.

The tablet I tried certainly took the bitter edge off licking a lemon, but the after-taste was unpleasant. An expensive red wine was transformed by it into something that was disgustingly sweet and fizzy. The only good thing was that the effects wore off within an hour.

Grasping the nettle

If you have a sweet tooth and are worried that you are

downing too much sugary stuff, I suspect 'cold turkey' is the way to go. But rather than rely on will power it's important to find something else to fill the void.

The Australian actress Rebel Wilson managed to lose 15kg by ditching sugary snacks and increasing the amount of fibre she ate. She also started doing 5-10 minutes of high intensity training (HIT) four times a week.

According to reliable sources (i.e. a celebrity website), Rebel aims for 35g of fibre a day, which she gets by eating oatmeal, fruits, veggies and wholegrains. When she feels a craving for sugar she has a shot of organic apple cider vinegar.

Rather than eating pretzels, bagels and bread, she now fills up on healthy fats such as olive oil and avocado. Her go-to snacks include:

Carrot and cucumber sticks dipped in guacamole
Almond butter on celery and carrots
A handful of almonds, brazil nuts or cashews

Like Rebel, I snack on nuts and veggies when the desire for cake or chocolate becomes almost unbearable. I also find it useful to imagine that I am slowly starving out those sugar-hungry microbes in my gut. Man versus Microbe. Surely there can only be one winner? If you want more advice about how and why you should cut down on sugar visit thebloodsugardiet.com

Processed foods

A couple of years ago Tim Spector, Professor of Genetic Epidemiology at King's College, London, persuaded his son Tom to go on go on a fast-food diet. For 10 days Tom ate nothing but the food he could buy from his local McDonald's. This included lots of Big Macs, chicken nuggets and fries, all washed down with Coke. Before, during and after, samples of Tom's poo were sent off to be analysed. Understandably, Tom did not feel great on this diet, and his gut biome had an even worse time. After a few days he had lost around 1,400 species, roughly 40 per cent of his total. Weeks after returning to his normal diet his biome still hadn't recovered.[36]

A few years ago I did a similar, albeit less extreme self-experiment. To see how bad a diet of processed food can be, I decided to go on one for four weeks – not living entirely on McDonald's like Tom, just eating processed meat once or twice a day.

On this new regime I started most mornings with a couple of bits of bacon. I then had a salad for lunch and a small burger for my evening meal. On another day, perhaps a sausage for breakfast and a salami sandwich for lunch. In many ways it was a fairly typical British diet.

To monitor the impact of this, before I started I went to the Food and Nutritional Sciences Department at Reading University. They gave me a thorough health check, measuring my cholesterol, weight, body fat and blood pressure and taking poo samples.

Four weeks later I returned.

When I stepped on the scales the nurse, with a slight smile, said I had put on an imprcssivc 3kg since my last visit, much of it around my gut. This was bad news because fat around your abdomen greatly increases your risk of becoming insulin-resistant and diabetic.

Then I had my blood pressure taken. Yet more bad news. I had gone from 'excellent' to 'stage 1 hypertensive'.

My gut bacteria had also changed, in a bad direction.

My diversity score (which is one of the better measures of general gut health) had fallen from 7.27 to 7.1 (where 10 is the most diverse). This doesn't sound much but it meant that whereas previously I had had a reassuringly diverse microbiome (I was in the top 30 per cent), now I had fallen right down the rankings and was languishing in the bottom 30 per cent.

There was also a major shift in my gut bacteria in favour of Firmicutes, the sort of bacteria that are linked to obesity and inflammation. After reverting to my normal diet my weight, blood pressure and biome gradually returned to normal.

It's not surprising that processed food is bad for you, but it is striking just how bad it is for your biome. One reason is the amount of sugar and fat in processed food, but there is also the matter of emulsifiers.

Emulsifiers, which are a sort of detergent, are added to most processed foods for texture and to extend their shelf life. Feeding common emulsifiers to mice tilts their biome in an unhealthy direction and encourages the growth of bacteria that attack the mucous lining of the gut. This in turn leads to inflammation, which contributes to the development of diseases like type 2 diabetes and obesity.[37]

Another good reason to say 'no' to burgers and fries.

Antibiotics

When I was a child we lived in India. While we were there I had multiple bouts of severe diarrhoea, which in turn meant numerous courses of antibiotics. These almost certainly saved my life, but may have contributed to the fact that I later developed type 2 diabetes (which I then overcame, see Chapter 6)

Many children are exposed to antibiotics before they are born. Pregnant women are routinely prescribed them, and they readily cross the placenta and reach the foetus. The mother of a baby who is going to be born by Caesarean section will also routinely get antibiotics, which will do nothing good for that baby's biome. The baby will already have a greater risk of developing asthma and type 1 diabetes than one born vaginally, as we saw in Chapter 4, and this early exposure to antibiotics may increase that risk.

The first two to three years of life are hugely important, not just for a growing child but also for their microbiome. That's when it gets established. Repeated doses of broad-spectrum antibiotics like penicillin, given in those first few years, significantly increases the risk of developing type 2 diabetes and becoming obese.

Some experts recommend that if a young child does have to have antibiotics, then it would be a good idea to take a course of probiotics at the same time, possibly with some fermented food like yoghurt or kefir as well. The same is true if you have to take a course of antibiotics as an adult. It's advice that Sophie Ramsbotham wishes someone had given her.

Sophie's story

Sophie is 29 years old and works as a garden designer. When she was in her early twenties she lived in India, designing class rooms for special-needs children. While she was there she got typhoid fever.

A prolonged course of antibiotics dealt with the immediate problem and she returned home, to the US. But over the next two years she struggled with terrible gut problems. These included pain, bloating and frequent diarrhoea. She was often bedridden and had to drop out of university, where she was studying architecture. She saw a lot of doctors.

'At least 20, and to be honest none of them had a clue of what to do,' she told me. 'They clearly felt pretty hopeless. They kept suggesting more tests, more invasive endoscopic procedures, more antibiotics. I was also told "You have IBS, go and deal with it".'

She ended up, reluctantly, visiting a neurologist, who

also described himself as a 'functional doctor'. He offered to heal her through diet.

'There were lots more blood tests. Lots of stool tests. He found plenty of nasties in my gut which he tried to kill with monolaurin [a substance derived from coconut that has antimicrobial properties].'

He also put her on a radical diet. His approach was to take out all the foods that might be a problem.

'I was down to chicken, sweet potato and broccoli. That was it. I was also taking lots of supplements to try and heal up the lining of my gut.'

The good news was that, for the first time in many years, her guts felt better. The bad news was, on her incredibly restrictive diet, she still felt run down, tired and lethargic most of the time.

She moved, with her husband, to London, where she found Tanya Borowski.

'Under Tanya's guidance I started to expand my diet again, eat things like lentils and grains. I don't think I will ever be back on dairy. These days we eat a lot of sauerkraut and kombucha which we make ourselves. My husband wants to get into kimchi. If I had to take antibiotics again I would also take high doses of probiotics and eat lots of fermented foods as well. No one tells you that. I wish they had.'

6

Other Ways to Improve Your Biome

Intermittent fasting

Back in 2012 I discovered, through a random blood test, that I was a type 2 diabetic. Rather than start on medication I decided to see if there was any way to reverse this. I spoke to lots of scientists and came across something I'd never heard about before, intermittent fasting.

Intermittent fasting doesn't mean cutting out food altogether. It means reducing your calories, fairly dramatically, a couple of days a week. Intrigued, I made a documentary called *Eat, Fast, Live Longer*, with myself as a guinea pig. As part of the documentary I decided to see what would happen if I cut my calorie intake to around 600 a day for two days a week, and did this for eight weeks.

I found it surprisingly easy to do, and on what I was now calling a 5:2 diet I lost 10kg, reversed my diabetes and took four inches off my waist. Based on my experience, and numerous conversations with experts, I wrote *The Fast Diet*, which became an international bestseller. This turned intermittent fasting into a worldwide phenomenon, and garnered fans including Jennifer Lopez and Benedict Cumberbatch. Five years on, I've maintained both the weight loss and the improved blood sugar levels.

So why does it work?

One reason why the 5:2 has been so successful is that you aren't on a constant treadmill, dieting all the time. It's easier to resist the temptation to eat something unhealthy if you say to yourself, 'I will have it tomorrow.' Then tomorrow comes and maybe you eat it. But often you don't.

Intermittent fasting also teaches you better ways of eating. If you satisfy your hunger on fasting days by eating vegetables and good protein, then over time you'll discover that when you get hungry you are more likely to crave the healthy stuff.

The science behind it is strong and getting stronger. A recent review article published in the scientific journal *Cell Metabolism*, 'Fasting: Molecular Mechanisms and Clinical Applications', looked at numerous human and animal studies and concluded that intermittent fasting reduces many of the things that encourage ageing ('oxidative damage and inflammation') while increasing the body's ability to protect and repair itself. According to this article it 'helps reduce obesity, hypertension, asthma, and rheumatoid arthritis.'

Fasting and the gut

If it is doing good things to your body, what effect is it having on your microbiome? Well, there is one species of bacteria we've met before that flourishes when we cut our calories or don't eat anything: *Akkermansia*.

Because *Akkermansia* lives on mucus and not on the remains of the food you've eaten, as most gut microbes do, it thrives when you reduce your calories. In the cut-throat world of the gut, *Akkermansia* has the upper hand in situations where its rivals are short of nourishment.

As well as animal studies, there have been a couple of small-scale human studies that have shown this. In one,[38] 13 overweight men and women were asked to go on a low-calorie diet for a week, eating 600-800 calories a day.

Poo samples were taken at the beginning and end of the week. The main finding was an increase in microbial diversity (which, as we've seen, is a good thing), and a dramatic rise in *Akkermansia* and *Bifidobacterium* (also one of the good guys).

In another small study[39] researchers from the American Gut project put 19 volunteers through a cheese and yoghurt diet, a dietary cleanse (fruit and vegetable smoothies or water with lemon, maple syrup and cayenne pepper) or fasting (water only or minimal calories). They took poo samples every day for the three days of the study and for a week afterwards.

Going on a fast was the most effective way to change the biome. It led to big increases in *Akkermansia*, which persisted for about 10 days after the fast had ended. Levels then slowly dropped back to where they had started.

So a short fasting blast can kick-start changes, but if you want to maintain those changes you may need to keep doing a bit of calorie restriction once a week or so (I call it 6:1, and that's what I do).

As well as being a Professor of Epidemiology, Dr Tim Spector, whom we met earlier, is a founder of the British Gut project and a leading microbiome researcher. Like me, he is a huge fan of the Mediterranean diet. 'It is the only one that has been proven to be beneficial for health,' he says. 'Nuts, seeds, dark chocolate, red wine, olive oil and vegetables like leeks, garlic and onions are packed with the chemicals that microbes love. We should also be eating more fermented foods like live yoghurt, kimchi, sauerkraut and kefir.'

Furthermore, he thinks that highly restrictive diets, which cut out whole food groups, such as Atkins, are likely to be bad for the biome. In fact, the only 'diet' he agrees with is intermittent fasting, preferably with a high-fibre intake. 'I like your 5:2 plan,' he told me, 'because it keeps food intake diverse.'

People often misinterpret the 5:2 approach, claiming that it allows you to eat 'whatever you like' on the days that you are not fasting. This is not true. The best way to do

5:2 is to go on a Mediterranean-style diet, reducing your calories two days a week. I'm not rigid about the number of calories for the fasting days. Some people find they get by comfortably on 600 calories. Others struggle. If you find that 600 calories leave you starving, then increase to 800 calories.

For more information on the science and how to do it, read *The Fast Diet* or visit thefastdiet.co.uk or theblood-sugardiet.com, where you will find a vibrant and supportive community of happy fasters.

Exercise

We know that exercise and increased activity are essential to keep your heart in good shape and your brain ticking over. More surprisingly, exercise also seems to be good for your biome.

Evidence for this comes from a study done by the Microbiome Institute in Cork, Ireland, where researchers decided to take a close look at the poo of the Irish rugby team.

Dr Paul Cotter, one the researchers, told me the reason they did the study with the rugby team was that they were an exceptional group of athletes. 'We thought that if this particular group of young men didn't have an altered gut microbiota then no one's gut microbiota would be altered by exercise.' So what did they find?

'The rugby guys had a weird microbiota, hugely diverse,' Paul said. 'We have only seen comparable results from people living in the Amazon rainforest or hunter-gatherers like the Hadza.'

This study provided the first real evidence that exercise increases microbial diversity in humans. Paul decided to follow it up by looking at the impact of exercise on people who don't usually do a lot of exercise.

'What we did,' Paul told me, 'was take a group of couch potatoes, and put them through an eight-week exercise regime. While they were doing the training we also gave them whey protein, because adding protein seems to be important. We took samples before, during and after, looking primarily for changes in their gut diversity.'

The study isn't yet fully complete, let alone published, but Paul is quietly excited. 'We didn't think we would see much in that short a time period; we thought it would merely be a stepping stone for a bigger study, but I've started crunching the data and we are already seeing some intriguing changes.'

Paul thinks that exercise will almost certainly help make your biome more healthy but he also told me, 'You can't outrun a bad diet.'

'As well as doing lots of exercise the Irish rugby players were eating a rich and varied diet. After training they would go to a special buffet where they ate lots of fruits, vegetables, white meat, nuts, berries, seeds, that sort of thing. The days of a beer and a pie after training are over for elite athletes.

So if you want to improve your biome, what type of

exercise and how long? The honest answer is that we don't know but it is likely that anything will help; running, walking, swimming, taking the stairs, even just standing around more will probably make a difference. Above all, avoid sitting around too much.

As someone who makes documentaries for a living, I regret having to tell you this, but television kills. Professor Spiegelhalter of Cambridge University has calculated that every two hours you spend sitting down watching TV cuts your life expectancy by about half an hour. The other problem is that if you are sitting down watching TV you are also probably eating crisps and other junk food.

One way round this is to watch TV standing up (perhaps doing the ironing, which burns about 80 calories an hour) or while on an exercise bike.

For optimal health you also need to inject some intensity into your exercise. The type of exercise I favour is high intensity training (HIT). For me this means very short bursts on a bike or just running up a flight of stairs, anywhere where I can really push myself.

This seems to be a particularly efficient way of building up the strength of your heart and lungs – even if you manage just a few minutes a week.

So how does HIT work? One of the things it does more effectively than low intensity exercise is stimulate your mitochondria – the body's main power plants. Their job is to convert raw materials such as oxygen and glucose into little packages of energy called adenosine triphosphate (ATP). The ATP is then used to power your body.

There are mitochondria in every cell of your body and, since they produce power, you want more of them. A good measure of how effective an exercise regime is going to be is whether it results in greater mitochondrial density.

That's where HIT scores particularly well. Doing HIT leads to greater numbers of more active mitochondria than standard exercise, and this is particularly impressive when it comes to heart muscle. HIT makes the heart bigger, stronger and more efficient. This is important not only if you are going for a run, but also if you are recovering from a heart attack.

In addition, mitochondria are crucial to fat-burning. When you start to exercise, your body breaks down the fat stores, releasing free fatty acids into your blood. These fatty acids are taken up by the muscle and converted by the mitochondria into energy. The more mitochondria you have, the greater the fat consumption.

Compared with people on standard exercise regimes, those who do HIT see a bigger loss in abdominal fat, which as well as being unsightly is a risk factor for diabetes and heart disease.

A few years ago I wrote a book called *Fast Exercise*, which extolled the virtues of HIT. Although people were enthusiastic about the science, there was also anxiety about the risks associated with pushing yourself hard, particularly if you are a bit older.

So I am delighted by a recent study, 'Extremely short duration high-intensity training substantially improves the physical function and self-reported health status of an elderly population', published in the *Journal of the American Geriatrics Society.*[40]

In this study, scientists from Abertay University in Scotland put a group of unfit, elderly volunteers through an exercise regime in which they were expected to do two minutes of HIT a week. The volunteers ranged in age between 61 and 74.

The volunteers, untrained individuals who had done no regular exercise over the previous year, were screened to make sure they weren't on any medication that would affect their performance. Then they were randomly allocated into an exercise or control group.

The two sessions consisted of 10 x 6 seconds of all-out sprints against resistance on an exercise bike, with at least one minute to recover between each sprint. They wore a heart monitor and were asked not to do another sprint until their heart rate had dropped back to below 120 beats per second.

The secret to doing HIT is not speed, but effort. You have to push yourself, which means doing it on something like an exercise bike, where you quickly increase resistance. As Dr John Babraj, the lead scientist put it, 'When it comes to the sprints, you don't have to go at the speed of someone like Usain Bolt. As long as you are putting in your maximal effort – whatever speed that happens to be – it will improve your health.'

By the end of the six-week trial, the volunteers doing their two minutes of HIT a week had lost an average of 1kg, mostly fat, despite having been asked not to change their eating habits. Dr Babraj thinks this is partly due to an afterburn effect you get from doing HIT, an increase in metabolic rate that persists for many hours. It is not much, around 70 calories a day, but over six weeks that would add up to around 2000 calories, the energy equivalent of 0.3kg of fat. Unlike conventional exercise, which tends to lead to compensatory overeating, HIT also seems to suppress appetite.

The real benefits of his two-minute exercise regime, for the more mature of us, were seen in its effect on blood pressure, aerobic fitness and blood glucose. High blood pressure is a significant risk factor for heart disease and stroke.

The volunteers doing HIT had an average fall in blood pressure of around 9 per cent, better than you would get with most drugs. They saw a similar drop in their blood glucose.

They also saw an 8 per cent improvement in aerobic fitness, which is a measure of the strength of your heart and lungs. Aerobic fitness is one of the most accurate predictors of how well and how long we will live.

Clearly people who are older need to take things more gently when they start, but there is a persistent and overstated fear that exercise in later life will lead to heart attacks and strokes. The reverse is true. However, if you are on medication or are worried about your health do consult your doctor before starting a new exercise regime.

The dangers of overdoing it

Although not doing any exercise is bad for you, there are dangers in overdoing things, even when you are fit. A lot of runners who really push themselves have gut problems, particularly after doing something strenuous like a marathon. I'm told, though I've never done one, that the biggest queues at these events are for the bathroom. 'Marathons are won and lost in the toilet,' I was informed by a friend. The problem is that excessive exercise can lead to gut damage and leaky gut syndrome. The damage to the gut is caused by an increase in body temperature and a reduction in blood flow. In most people it seems to be only a temporary effect, but just be aware that though modest levels of exercise will improve your gut, it won't enjoy being pushed too far.

Stress, sleep and the biome

One of the things that exercise is good for is relieving stress. Too much stress, lack of sleep and an unhealthy biome are all interlinked.

Stress makes you sleep badly, which makes you eat badly (it cranks up your desire for sugary carbs and high-fat snacks), which encourages the growth of 'bad' microbes in your gut. You put on weight, get more cranky, sleep even worse and so on and so on. It is a horrible, vicious cycle. The immediate effects of a bad night's sleep are particularly striking when it comes to appetite.

In one study in which they asked 27 men and women (aged 30-45) to cut down their sleep to four hours for just one night, they found dramatic changes in their hunger hormones and appetite.[41] Oddly enough, sleep deprivation affected men and women differently. In the case of the men, the reduced sleep led to a surge in the hormone ghrelin, which tells you that you are hungry.

In the women that didn't happen. Instead, there was a fall in a hormone called GLP-1, which tells you when you are full. The effect was the same, the subjects ate more, but for different reasons. It's not clear why there is this gender split.

Shift work and jet lag also do pretty awful things to your body and your biome. On a recent filming trip I got chatting with an air stewardess who told me she survives on sleeping pills and energy drinks. She knows this is a terrible combination, but she's in a rut. 'I haven't told my husband,' she confided. 'He thinks I'm taking iron tablets last thing at night.'

The Erans (Dr Elinav and Dr Segal), the brains behind the Personalised Nutrition project, decided to explore this and asked some students to fly across time zones. Naturally enough, they were also asked to collect their own faeces samples before they got on the plane, later when their jet lag was particularly bad and then two weeks after that.

The Erans then transferred the bacteria from the students into germ-free mice. Dr Elinav says the results were startling. 'The mice that received the microbes from jet-lagged students grew obese and developed diabetes. Those that were given bacteria from samples taken before or after the jet lag set in were unaffected.'

The moral is, try to avoid flying across lots of time zones, if you can. The other moral is prioritise a good night's sleep. Easier said than done.

If you are sleep deprived, what can you do about it?

There's plenty of sleep advice out there, most of which I've tried and some of which has helped. The best-tested, scientifically based advice includes the following:

1. Set your alarm clock to wake you up at the same time every morning. This creates a ritual and your body loves rituals. Sadly, that means no weekend lie-ins.

2. Try to go for a walk before breakfast. Morning light switches off your brain's production of melatonin, the hormone that helps put you to sleep. Bright light really will perk you up. If you suffer from seasonal affective disorder ('the Winter Blues') this will be particularly helpful.

3. Do some exercise during the day. Lots of studies show that exercise improves quality of sleep.

4. Prepare for sleep at least 90 minutes before you go to bed. Eating late at night is really bad for your

body and forcing your gut to do lots of heavy digestive work is not going to improve the quality of your sleep.

5. Having a warm bath or hot shower at least an hour before going to bed can also be helpful, as long you cool down afterwards. It is the drop in body temperature that cues your brain that it is time to sleep. That is also why you should try to sleep in a cool room.

6. Avoid watching TV, being on your phone or social media in the run-up to bedtime. This is partly because these activities are overly stimulating, but also because the blue light screens typically emit will wake your brain up. Do not, whatever you do, watch TV or use your laptop in bed. Bed should be for sleeping and sex, nothing else.

7. Cut down on late-night boozing. Alcohol may help you go to sleep but it will also ruin your REM sleep. Whether or not you choose to drink coffee in the evening is a more individual thing. Some people are fast caffeine metabolisers, meaning they have a gene that breaks down caffeine particularly fast. I've been tested; I'm a fast metaboliser and drinking coffee at night makes very little difference to my sleep quality. My wife, on the other hand, is a slow metaboliser and can't drink coffee

after lunch or she will sleep really badly.

8. Before you go to sleep, write down three good things that happened to you that day. This has been shown to be one of the best ways to improve mood.

That's the well-established stuff, which if you are an insomniac you have probably already tried. You may also have tried lavender or other herbal remedies. The evidence for these is weak, but they may work for you.

Now let's move on to one of the more surprising approaches to improving your sleep patterns: changing your biome.

Fibre and sleep

Eating fibre, as we've seen, increases levels of 'good' bacteria, such as *Bifidobacteria*, which in turn produce all sorts of hormones and metabolites that improve not only your gut health, and your weight but also, possibly, your brain. There is now evidence that more fibre will help you sleep better too.

In a recent study researchers from the Institute of Human Nutrition at Columbia University Medical Center in New York brought 13 men and 13 women, average age 35, into their sleep clinic and tested different diets on them. These included changing the amounts of fibre, saturated fat and sugar in the meals they were eating.[42]

What they found was that having meals rich in fibre improved the quality of the volunteers' deep sleep and also meant they got to sleep faster (down from an average of 29 minutes to 17 minutes). Eating more saturated fat and sugar, on the other hand, led to poorer quality sleep – it was lighter and less refreshing.

As Dr St-Onge, one of the main researchers, pointed out, 'The finding that diet can influence sleep has tremendous health implications, given the increasing recognition of the role of sleep in the development of chronic disorders such as hypertension, diabetes and cardiovascular disease.'

Potato starch and sleep

As well as cramming in more veg, I've been trying a type of fibre that *Bifidobacteria* seem to particularly like – potato starch. Potato starch is a form of resistant starch, made by crushing uncooked potatoes and then harvesting the starch grains from the destroyed cells. You can buy it in health food shops or online. Although there have been no clinical trials that I know of, the anecdotal evidence (mainly internet chatter) is that a teaspoon taken in the early evening helps sleep and also leads to more vivid dreams.

If anyone likes the idea of testing this out and contributing to science, I would really appreciate it if you go to our website and sign up for a small study. It will be online and completely anonymous, but with your help we may be able to find out how useful, if at all, eating potato starch really is. If you would like to take part it is important that you read the protocol on the website before beginning, so we can be sure that the results are valid. Many thanks in advance.

Stress

Sleep deprivation leads to stress and stress leads to sleep deprivation. Along with 'eat less and do more exercise', one of the more annoying bits of advice that well-intentioned people like to dish out is 'try to relax'. We all know we should relax, that too much stress is bad for us and that it's spoiling our lives. The real question, beyond improving sleep quality, is what can we do about it?

I've already covered probiotics in an earlier chapter, and we know that more exercise will help. But what about other, more psychologically based techniques?

One of the tricks I find helpful when I am in an acute stressful situation, like preparing for a radio interview or doing a live TV show, is to try telling myself that what I am experiencing is not stress but excitement.

In both cases you get a rush of adrenaline, sweaty palms, agitation. Your body knows that you are not relaxed, but what it can't tell is whether you are excited or stressed.

So, repeating to yourself, 'I'm really looking forward to this, which is why my pulse is racing and my hands shaking' may help. It is certainly better than thinking, 'This will be a disaster, I wish I wasn't here, help, help, help.'

For more chronic stress I recommend mindfulness, a subject I've written about before but is worth touching on here.

Can you change your mind?

Do you ever find yourself going for a long drive and

reaching the end without being aware that you have been driving, lost as you are in your own musings?

Most of us spend our waking lives wrapped up in our own internal world. We over-think and, like overdoing anything, over-thinking tends to have negative consequences. In the case of constant mental meanderings the risk is that they will lead to a negative spiral of indecisiveness, self-loathing, depression and insomnia. One way to counter this is to make yourself more mindful.

I had been intending to try mindful meditation for some years, but never quite got round to it. Finally, three years ago I decided it was time to sort out my head. Because I wanted to measure, objectively, what effects it would have on my brain, before starting I went to see Professor Elaine Fox, then at the University of Essex, now at Oxford University.

She did a range of tests, including measuring the levels of electrical activity on the two sides of my brain while I was resting. Surprisingly enough, studies have shown that people who are prone to high levels of pessimism, neuroticism and anxiety tend to have greater activity on the right side of their frontal cortex than on the left. This is known as cerebral asymmetry. We know it happens but we don't know why.

The results of the various tests suggested that I was a pessimist with a brain which, as Elaine kindly put it, 'is on the negative side of the spectrum'.

Then I went off to try mindful meditation for six weeks. Some people like joining a group, but I was running all over the place so I ended up doing it via an app.

If you want to get a flavour, try this: sit up straight in a comfortable chair, rest your hands on your thighs, close your eyes and then for the minute or so try paying attention just to your breathing.

You don't need to speed it up or slow it down, simply pay attention to the sensation of the breath going through your nostrils, filling your chest, expanding and contracting your diaphragm. When you notice that your thoughts have drifted, which they will, gently bring your focus back to your breath.

OK, don't read any more. Don't turn over the page. Put this book aside and give it a go. It's only a minute, what have you got to lose?

... So how did it go? I predict that doing it for even a minute was tough. You probably felt uncomfortable, perhaps experienced an inner resistance, a feeling that this is a complete waste of time. This is absolutely normal, a sign that your thoughts, like wild horses, are not used to being controlled.

The Raisin Test

Another way to experience 'mindfulness' is to do the raisin test. You can also do this with a square of chocolate, but I find that too distracting. Start by reading these instructions, then put the book aside. Or do it with a friend who is happy to read them out to you. Or go to our website where you can listen to me guiding you through it.

1. Take a raisin, just one, and put it in the palm of your hand. Take a close look at it. Inspect it as if you have never seen a raisin before. Pay attention to the colour, the weight, the way the light reflects off its surface. Spend some time admiring it.

2. Next, pick it up with you other hand and give it a little squeeze. Roll it around in your fingers. Enjoy the texture.

3. Now give it a sniff. What does it smell like? What does it remind you of? Hot summer days? A glass of wine on holiday? Being young and happy? Close your eyes, enjoy the moment, breathe in those raisiny smells, deeply.

4. Now put it on your tongue, but don't chew or swallow it. You can use your tongue to roll it around, but you have to resist biting. Again, close your eyes so you can properly savour it.

5. Finally, bite into it, slowly, and notice the release of flavours.

6. When you are ready, swallow it, but try to stay in the moment, simply enjoying the lingering effects of the now vanished raisin on your taste buds.

When I started doing mindfulness I spent ages finding excuses not to do it. Too busy, too tired, too hungry, too stressed. The thought of Elaine Fox waiting for me at the end of the six weeks kept me on track. I also found it helpful that my wife, Clare, joined in. Even so, I spent much of the allotted time (initially 10 minutes a day, building up to 20) absorbed in my usual concerns, having to be constantly nudged back on course by the voice on my app.

The good news is that like any form of exercise it got easier to do, though I rarely managed more than a few minutes of focus at a time.

As well as sitting quietly, I tried building mindful moments into my day. Instead of just knocking back a cup of coffee, I would hold it, feel the warmth, try to focus on the muscle activity involved in bringing it to my lips, feel the warm liquid cascade down my throat, etc.

At the end of six weeks I felt noticeably calmer and my insomnia was also more under control. But what about my brain? I went back to Elaine to be retested and was amazed by the results. Whereas before my brain activity had suggested I was a negative pessimist, now there was a much better balance of activity between my two hemispheres, indicating a sharp reduction in negative thoughts and emotions.

But why does mindfulness have this effect? Probably because it helps strengthen your control over your own thoughts and feelings. In a study published in the journal *Social Cognitive and Affective Neuroscience,*[43] they asked 15 volunteers to do four sets of 20-minute classes of mindfulness. Then they used a special type of brain scan, arterial spin labelling magnetic resonance imaging, to look at their brains. The volunteers also filled in an anxiety questionnaire, before and after.

The researchers found that mindfulness reduced anxiety ratings by up to 39 per cent. They also found that it increased activity in the areas of the brain that control worrying, particularly the ventromedial prefrontal cortex and the anterior cingulate gyrus. This supports the claim that mindfulness strengthens our ability to ignore negative thoughts and feelings.

Conclusion

I have become, as you can probably tell, a little bit obsessed by guts in general and the microbiome in particular. Talking to researchers and reading about the latest developments has been an incredibly exciting experience. It feels like we have come across this brand-new land, populated by strange and exotic creatures, which are only now beginning to reveal their secrets.

As with any new discovery, there are exploiters as well as explorers. Peddlers of dodgy supplements and useless probiotics are well aware that there is a huge market out there, which they are keen to make money from. Food manufacturers, having sold us the alleged benefits of low-fat foods, are gearing up to produce a wide range of biome-friendly foods. Go carefully, is my advice. Unless you have a specific reason for doing so (such as IBS or to counteract antibiotic-related problems), buying probiotics may be a waste of time and money.

Instead, where possible, eat unprocessed 'real' foods. Explore fermented foods. Use the recipes in this book to expand your repertoire and give your biome something to chew on. Good food should be a pleasure and it should be shared. Not just with your friends but with your Old Friends.

Keeping your biome properly fed and cared for is definitely worthwhile. If you look after all those friendly microbes then they will look after you. We are at the start of what I am convinced is a whole new way of approaching and understanding nutrition, one that could change the way we treat a wide range of diseases, from obesity to depression. This is just the beginning – there is so much more to come.

PART II

How to Reboot Your Biome

a 2-Stage Healing Programme

If you have mild IBS or problems with abdominal pain and bloating, you may want to reboot your gut bacteria. The principle is fairly straight forward – if your symptoms improve or disappear when you remove a certain food item and then return when you re-introduce it, the symptoms are likely to be dietary. If, however, there is no change as a result of exclusion and reintroduction of a particular food, it is unlikely to be the cause of your symptoms.

The following is a simple programme that many people have found useful. There are two phases: a remove and repair phase and a reintroduction phase. This is R & R for your gut. Ideally you should spend up to four weeks on the repair phase, though some people experience such rapid improvement they only need to it for two weeks. The length of the reintroduction phase depends on whether or not you encounter problems when you reintroduce particular foods. The message is, go slowly and carefully.

Before you start

- If you have any significant medical problems or very troublesome symptoms, we recommend that you consult a health professional first. There may be medical reasons for your symptoms that need checking first, such as undiagnosed coeliac disease, which it is important to have investigated. Undiagnosed, coeliac disease can cause low-grade symptoms or even no symptoms at all, but it can still lead to significant complications.

  Other conditions such as inflammatory bowel disease may require further investigation or management.

  It is important to rule out other potentially significant causes of your symptoms before embarking on dietary changes and to identify any reasons why this might not be a suitable approach for you. It may help to do the programme with professional support.

  If you are underweight, suspect you have a food allergy, have other significant medical problems or are frail or unwell, we would not advise embarking on this programme.

- Keep a detailed daily Food and Symptoms Diary for at least a week before you start, to help identify any pattern in relation to diet and other factors (see example on page 266 or print one via resources at cleverguts.com). Ideally keep the diary going throughout the programme until you identify any culprits. Although it requires an effort to maintain, a diary will help you be systematic and focused in your approach. If you see a professional, you will also have useful information available.

- Plan ahead. Go to a good-sized supermarket to identify what options there are to 'replace' a missing food. This may be in the 'free from' area – for example, the gluten-free products such as breads and oats. Often you can find natural substitutes, for example rice or quinoa for gluten-containing pasta.

  If you have any significant events coming up, or holidays that will make it hard to sustain, defer your starting date.

Helpful practices to incorporate throughout:

Food-related:

- Cut right down on sugars, refined starchy carbohydrates, processed foods and sweeteners, including sweetened soft drinks and juices. These are a disaster for gut bacteria and inflammation.

  We recommend you continue to keep these to a minimum during and after the programme for continuing gut health.

- Avoid trans and partially hydrogenated fats. These are mainly present in spreads and some processed foods including biscuits and pastries.

- Practise intermittent fasting – 12-14-hour overnight fasts or 5:2 fasting (5 days eating normally, 2 days low-calorie fasting). Giving your gut a rest from having to constantly digest food allows the lining to regenerate and encourages the growth of good bacteria like *Akkermansia* (see page 56).

General strategies:

- Sit down for meals and chew your food thoroughly – try to make your mealtime relaxing! It might sound obvious, but chewing thoroughly reduces the likelihood you will overeat, because it gives your gut more time to release 'I'm full' signals.

- Lower your stress levels – staying calm soothes the gut and reduces levels of hormones such as cortisol that upset the balance of the biome.

- Improve your sleep. It is vital to get enough quality sleep, which means not eating meals late at night and committing to being in bed by 10.30pm at least four nights a week.

- If you are constipated try magnesium citrate or triphala, a traditional Ayurvedic herbal treatment. You may also want to try Pia's seaweed capsules (see page 133). And drink plenty of water.

- Open the windows, get outside, start gardening. You need to let more bugs into your life.

Phase 1 – REMOVE & REPAIR

Depending on how bad things are, you may need at least four weeks of R & R to give the lining of your gut a chance to repair and recover. Foods which commonly cause gut problems include gluten, dairy, eggs, soya and coffee. You may also have particular foods that you suspect are causing a problem.

We don't recommend removing too many foods at one time, so it might be helpful to do 'remove and repair' in several stages. You can always repeat the process with the less troublesome foods at a later date.

During the Remove and Repair phase try to avoid:

- Gluten and refined grains – if you remove these temporarily from your diet it will also leave more room on your plate for vegetables to boost your gut flora.

- Dairy products – particularly milk, as this contains the most lactose. The process of fermentation reduces the lactose load in yoghurt and cheese. Though you may decide to remove all dairy products.

- Pulses – these contain lectin, which may contribute to bloating (reintroduce these after two weeks, particularly if you are a vegetarian, as they are a good source of protein).

- Alcohol (sorry).

- Very fibrous vegetables, such as broccoli and kale stalks, stringy beans, whole peas and brassicas such as cabbage and brussel sprouts. These tend to be easier to digest if they are slow-cooked. They can be reintroduced in the second phase.

Do include plenty of:

- Non-fibrous, plant-based foods – aim to fill at least half your plate with vegetables, herbs and fruit. Include at least seven portions a day of veg and fruit, made up mainly of vegetables. And make them colourful. Variety is important for gut health, so try to eat 20-30 different varieties a week.

- Good-quality proteins – these are necessary for the repair of your gut lining. You should aim for at least

45-60g protein a day. Choose from fish (particularly oily fish), eggs, chicken, game, red meat (grass-fed), soya, nuts, tofu, tempeh, etc.

- Bitter leaf salads with vinegar or a citrus fruit-based starter – these will help to get your digestive juices flowing.

- Polyphenol-rich foods, such as herbs, spices, nuts, seeds, fruit and berries, teas, red wine and dark chocolate (yess!).

- Phytonutrients – as in non-starchy vegetables and fruits of different colours, though restrict tropical fruits, melon and grapes because of their high sugar content.

- Non-dairy fats – such as olive oil, coconut oil, avocado, nuts and seeds.

Phase 2 – REINTRODUCTION

Once you have been through the Remove & Repair phase and are hopefully beginning to feel better, you can start to introduce new foods and reintroduce the ones you eliminated during the repair phase.

You can also start boosting your good bacteria by including more prebiotic and probiotic foods.

Introduce foods one at a time with a gap of at least three days between each one. The point is to try and identify the foods that are causing you problems.

Increase prebiotics

These are the foods that feed the good bacteria:

- Jerusalem artichokes, onions, leeks, onions, garlic, fennel, asparagus, apples, chicory, bok choy.
- Pulses – these can be added back in after two weeks. Pulses are an important protein source for vegetarians, but reintroduce these slowly as they can cause wind and bloating.

Increase probiotics

These are foods that help you top up your 'good' microbes:

- Fermented vegetables, including sauerkraut.
- Live yoghurts.
- Kefir.

- Cheese – particularly smelly cheeses like Roquefort, which are rich in bacteria.

Reintroduce excluded foods

These are to be added one at a time, over three days. Eat a normal portion of the suspected food. If symptoms recur after reintroduction, usually over the following few days, then withdraw it, allow a few days of recovery then try reintroducing something else.

Use the Food and Symptoms Diary (see page 266) to track your response.

- Dairy products – start with full-fat live yoghurt, cheese and butter, then add milk.

- Wheat/gluten – start with relatively low-gluten grains such as rye, spelt or flaxseeds. Sourdough breads are usually easier to digest as the lactic acid in the fermentation process helps neutralise the phytates in the flour. Then introduce wheat, again over a few days.

- Alcohol in moderation – and with food. Choose red wine if you can.

Important considerations

If symptoms recur after completion of this plan, you can hit the reset button and repeat the whole process. However, if your symptoms are particularly troublesome and not settling as expected, we recommend you see a health professional trained in this area to give more tailored advice and perhaps investigate further.

Seek more urgent medical advice if you:

- Are passing blood and/or mucus
- Have severe and/or persistent abdominal pain
- Experience unexplained, unplanned weight loss or loss of appetite
- Have a recent change in bowel habit
- Suffer from anaemia or a deficiency in important vitamins or nutrients
- Have persistent diarrhoea and/or vomiting

If you suspect you suffer from a food allergy, it is very important that you see your doctor and get tested as it could be life-threatening. Fortunately, food allergies are relatively uncommon. Symptoms normally occur within minutes of being in contact with or eating the relevant substance. The typical reaction might involve a blotchy red rash, which is raised and itchy, known as urticaria. There may be vomiting and/or severe gut symptoms such as diarrhoea; respiratory symptoms resulting in wheezing and difficulty breathing; itching or swelling of the lips, tongue and palate; or, very rarely, sudden collapse.

Food allergy occurs when the body mistakenly reacts to food substances by producing an antibody, IgE, to fight it off, when eaten (or rarely even when in contact with the skin). IgE can be measured by a blood test. Once your health professional has helped identify the food substance, you can take measures to avoid it and, if needed, keep emergency medication to hand. An allergy is different from a food intolerance, which is a non-allergic hypersensitivity, and is much more common. With an intolerance, the onset tends to be delayed by hours, not minutes, and the symptoms are more variable.

If you find yourself excluding food or food groups for a long period of time, we recommend that you consult a professional to ensure that the full clinical picture is considered and that you are getting a healthy balanced diet.

This approach is not suitable for children.

RECIPES

by Tanya Borowski, mBANT, IFMCP, and Dr Clare Bailey, GP

To start the day...

Standard breakfasts have become very centred around grains in the form of processed cereals and toast. Grains are a source of carbohydrates which can cause flatulence and bloating. We recommend sticking to wholegrains while following our Clever Guts plan, and eating less of them in general.

Pumpkin Porridge
(serves 2)

To put a warm glow in your belly and set you up for the day, why not try our Pumpkin Porridge? Pumpkins are high in beta-carotene, a powerful antioxidant, and in vitamin A (which supports the gut lining), C and E. They are also rich in fibre, which boosts our microbiome diversity.

300g pumpkin (or butternut squash), cooked and diced
50g coconut butter (or 30g coconut oil)
225ml full-fat coconut milk
1 tsp vanilla bean paste (or 1 tsp vanilla extract)
Pinch of salt
½ tsp ground cinnamon
Handful of pomegranate seeds or blueberries, to serve

Blend all the ingredients except the fruit until you have a smooth paste. Warm the mixture gently in a pan. You may like to thin it with a little extra coconut milk.

Serve it warm with the pomegranate seeds or blueberries.

Nutty Cinnamon Granola
(makes 10 portions)

This is one of Michael's favourites. It's a delicious combination of nuts, flaxseeds and wholegrains that will keep you full for longer. It's much tastier and lower in sugar than the shop-bought versions. Have it with some homemade yoghurt or kefir (see page 206), and for a bit of juicy sweetness add an unpeeled diced pear.

100g coconut oil
100g honey or maple syrup (add extra 20-30g if you have a sweet tooth)

2 tsp ground cinnamon
1 tbsp vanilla extract
1 egg white, whisked
100g rolled oats, rice flakes or buckwheat flakes (or a combination of all 3)
400g mixed nuts (choose from walnuts, almonds, hazelnuts, pecans or cashews), chopped
80g ground flaxseed
1 tsp sea salt

Preheat the oven to 100°C. Gently heat the oil, honey and cinnamon in a saucepan until the honey melts. Remove the pan from the heat and stir in the vanilla extract.

Allow the mixture to cool before mixing in the egg white.

Meanwhile, combine the oats, nuts and flaxseed in a large bowl, then stir in the contents of the pan. Place the mixture in small clumps on a baking sheet and bake it in the oven until it's golden.

This should take 1-1½ hours. Check it regularly and turn it once or twice. To ensure your granola is really crisp, you can then turn the oven off and leave it for a few hours or even overnight.

Allow it to cool thoroughly before storing it in an air-tight container. Serve it with live organic Greek yoghurt and berries of your choice or a diced pear.

Creamy Cashew & Banana Breakfast Pot

(serves 2)

Bananas are a prebiotic and encourage the growth of those health-giving gut bacteria.

250g raw cashews
1 medium frozen banana
200ml unsweetened almond milk
½ tsp ground cinnamon
2 tsp nut butter
1 tsp vanilla extract

Place the cashews in a bowl and cover them with water. Soak them overnight or for at least 6 hours, then drain them and blend them with all the other ingredients at a high speed, until you have a smooth paste. Decant the mixture into 2 pots and place them in the fridge. Serve them cold.

Tip: Keep some peeled, diced bananas in a zip-lock bag in the freezer.

Chia Pots

(serves 1)

2 tbsp chia seeds

125ml dairy-free milk of your choice (e.g. coconut, hazelnut or almond milk)

½ tsp vanilla extract

½ tsp ground cinnamon

1 tbsp fresh or frozen berries

Put all the ingredients except the berries in an airtight, or covered jar (approx 300g volume) and stir well. Add half the berries and stir again. Put the jar in the fridge, ideally overnight but for at least 30 minutes. The seeds will have absorbed the liquid and plumped up. Serve them with the remaining berries and one of these additional toppings:

- 1-2 tbsp toasted coconut flakes
- 1 tbsp ground almonds or flaxseed
- 1-2 tbsp blueberries or raspberries
- 1-2 tbsp whey protein powder (a great source of protein)

Healthy Gut Green Smoothie

(serves 1)

We don't like shop-bought smoothies as they are usually way too sweet, but we love this homemade one. Spinach is rich in vitamins and flavonoids, which have been shown to boost beneficial bacteria in the gut, while avocado is packed with good natural fats and vitamin E.

2 handfuls of organic spinach leaves
220ml water
½ avocado
1 medium banana
1 tbsp tahini
1 tbsp root ginger, chopped (optional)
Juice of 1 lemon

Blitz all the ingredients together in a blender until thick and creamy.

Kiwi & Chia Seed Smoothie

(serves 1)

Perhaps we should call this one 'The Unblocker'. It should certainly help move things along if you are prone to constipation.

1 kiwi

¼ medium avocado

4 tbsp chia seeds

Juice of ½ lime

220ml water

Blitz all the ingredients together in a blender until smooth.

Tanya's Leaky Gut Healing Smoothie

(serves 1)

This is one that Tanya often recommends to her patients. The collagen provides extra protein, and is particularly helpful for those with a leaky gut. A 2014 study showed consuming collagen powder can improve skin elasticity too.[44]

200ml unsweetened coconut or almond milk

1 block of frozen spinach or a handful of fresh spinach leaves

50g frozen berries (e.g. raspberries or blueberries)

2 tbsp collagen powder, organic (optional)

1 tsp ground cinnamon

Blitz all the ingredients briefly on a high speed in a blender.

Nut Milk

(makes 1 cup)

Dairy foods, with the exception of homemade yoghurt (if tolerated), are best avoided if you have gut problems as the casein can trigger inflammation and irritate the gut lining, while lactose can cause bloating and diarrhoea. You can buy non-dairy milks such as almond, coconut or hazelnut – or make your own, following this simple recipe. It produces a luxurious, nutritious, creamy milk, plus you know exactly what's in it.

125g raw unsalted cashews
150ml filtered or bottled water
Small pinch of sea salt
½ tsp vanilla bean paste (or ¼ tsp vanilla extract)

Soak the nuts overnight, or for at least 4 hours, then drain and rinse them. Put them in a blender with the rest of the ingredients and blend at a high speed until they have broken down.

To create a smooth texture, strain the mixture into a bowl through a piece of muslin or fine sieve. Squeeze out the fluid and discard the contents of the muslin. Store the milk in an airtight container for up to 4 days. Serve it chilled or heat it gently to make a creamy hot drink.

Turmeric Latte
(serves 2)

Turmeric, like ginger, is a natural anti-inflammatory. Combined here with cardamom and ginger, it makes a warming, aromatic drink.

1cm root ginger (or ½ tsp ground ginger)
2cm root turmeric (or 1 tsp ground turmeric)
300ml almond or coconut milk (from a carton not a tin)
Seeds of 2 cardamom pods
1 tsp local honey
2 tsp coconut oil
1 pinch ground cinnamon

Peel and grate the fresh ginger and turmeric, if using. Gently heat the nut milk in a small pan, then add the cardamom seeds, honey and coconut oil, whisking constantly so the milk heats through and is foamy. Pour the mixture into a cup and sprinkle over the cinnamon.

Homemade Yoghurt

Not only does homemade yoghurt usually have a higher probiotic content than shop-bought versions, it's also much tastier. And if you get into a routine, it just keeps on producing.

When fermented for 24 hours, most of the lactose is removed, which means that some people who are usually intolerant to lactose may well be able to 'tolerate' it. The easiest way to make your own yoghurt is probably with a yoghurt maker, although there are plenty of people who do it in the traditional way, keeping it in a warm place, covered, overnight. Some store it overnight in the microwave, which acts like a large thermos flask keeping it warm.

1 litre organic full-fat milk (or half and half of full-fat milk and single cream for a creamier texture)
2 heaped tbsp good-quality live organic yoghurt, or 2 tbsp yoghurt starter
1 litre thermos flask, or a covered bowl or a yoghurt maker
Digital thermometer (optional)
1 litre glass containers

Make sure all your equipment and containers are meticulously clean and have been sterilised in boiling water (or on a hot cycle in the dishwasher).

If using live yoghurt as a starter, it helps if it is at room temperature. The jars for storing the yoghurt should also ideally be at room temperature.

Put the milk in a pan and heat it gently, stirring frequently so it doesn't burn. Take it off the heat just before it comes to the boil (when it starts to bubble at the sides). Allow the milk to cool to about 30-40°C – it should feel just warm to the touch. (If using cream, stir it in earlier as this will also aid the cooling-off process). Then pour a cupful of milk into a bowl and stir in the starter. Add the rest of the milk and stir again so that they are well combined.

For incubation, transfer the mixture to a yoghurt maker, a dry sterilised thermos flask or a covered bowl and let it stand on the kitchen surface for 24 hours (many yoghurt makers can't be set for longer than 15 hours, so just reset again when the 15 hours are up if needed). The warmer the temperature, the faster it will set (it should never be allowed to get any warmer than 46°C).

Transfer the yoghurt to clean glass containers and store them in the fridge (to stop fermentation) for up to 5 days. Keep 4 tbsp yoghurt to seed the next batch within the 5 days and just keep the production line going…

Tip: Use goat's or sheep's milk if you are sensitive to dairy. They contain a different type of casein, which you may be able to tolerate.

Milk Kefir
(makes 500ml)

Kefir has been an enjoyable discovery. It is a fermented milk drink with a delicate flavour, and is easy to make. As with yoghurt, fermentation breaks down the lactose. Kefir culture is grown from a complex ecosystem of 'grains' that look a bit like tiny, soft cauliflower florets but are in fact living communities of around 40-50 types of bacteria and yeasts. Together they produce one of the most probiotic-rich drinks available.

1-2 sachets of kefir starter culture powder or 2-3 tbsp fresh grains
1 litre organic full-fat milk
1 litre glass container with lid

Follow the instructions for Homemade Yoghurt for sterilising your equipment.

If using powder, mix it into a smooth paste with a little milk in a jug or bowl, then add this to the rest of the milk in a glass container and stir. If using grains, simply drop them into the bottom of the container, add the milk and stir briefly. Place it in a warm spot for about 24 hours – the ideal temperature is 22-24°C. The kefir sets quicker in warmer temperatures and can take up to 30 hours on a cold day. Don't stir it during the fermentation process.

It is ready when it is lightly set; you can test by scooping out a teaspoonful – it should leave a small indent in the surface. Strain it through a fine nylon sieve or muslin and place it in a covered glass container in the fridge to cool. The grains can be used again – if not using them immediately, rinse them in filtered water and freeze. In fact, if you get into the flow, you will soon have more grains than you need – give some away to create another colony!

For variations in flavour you might add a spoonful of chopped fruit, a few drops of vanilla extract or a fruit-flavoured tea bag. Kefir is delicious served with granola (page 198) and fruit.

Raspberry Chia Jam

A delicious fruity low-sugar jam which is wonderfully quick and easy to make. You can also serve it as a coulis, stirred into full-fat yoghurt or coconut-based yoghurt.

30-40g pitted dates, finely diced (or 1-2 tsp maple syrup)
125g raspberries (frozen or fresh)
2-3 tsp chia seeds (depending on how thick you like it)

Gently heat the dates in a small pan with 2 tbsp water for about 2 minutes, stirring to form a smooth paste. Add the raspberries and chia seeds and simmer for 2-3 minutes. Mash the raspberries with a potato masher or the back of a spoon. The jam can be stored in an airtight container in the fridge for up to a week.

Green Flaxseed Bread

(makes a 900g loaf, 10-12 slices)

Flaxseed gives this bread a delicious, nutty flavour. Not only that, it is also rich in omega 3 and fibre, which your biome and your bowels will enjoy.

80g ground almonds
200g ground flaxseed
50g chia seeds
50g pumpkin seeds
4 eggs
200g fresh (or defrosted frozen) spinach
50g coconut oil or butter
1 tsp bicarbonate of soda
2 tbsp lemon juice
50ml water
1 tsp sea salt

Preheat the oven to 170°C. Line a loaf tin with parchment or grease it with coconut oil. Put all the ingredients in a food processor and blend until you have a smooth dough. Press it into the prepared tin and bake it for 45 minutes until is cooked through (if you pierce it with a skewer it should come out out clean). This is a fairly dense bread. Leave it to cool, then turn it out and cut it into slices. Store it in a zip-lock bag in the fridge or in the freezer. It's delicious toasted.

Breakfast Bread

(makes a 900g loaf, 10-12 slices)

The ground almonds, flaxseed and eggs make this a high-protein, low-grain breakfast or snack. This bread is best served warm or toasted. Try it spread with a little nut butter.

- 190g ground almonds
- 2 tbsp coconut flour
- 40g ground flaxseed
- 1 tsp sea salt
- 1½ tsp baking powder
- 5 eggs
- 30g coconut oil
- 1 tbsp maple syrup
- 1 tbsp live (raw) apple cider vinegar

Preheat the oven to 180°C. Line a loaf tin with parchment or grease it with coconut oil. Put the ground almonds, coconut flour, ground flaxseed, salt and baking powder in a food processor and pulse until everything is combined. Then add the rest of the ingredients and pulse again to form a smooth dough. Bake it for about 40 minutes or until it's golden and cooked through (if you pierce it with a skewer it should come out clean). Leave it to cool, then turn it out. Store it in a zip-lock bag in the fridge or in the freezer, sliced.

Lunchtime

Before a meal, to get your digestive juices going, why not have a small 'gin and tonic': a glass of fizzy or still water with a tablespoon of live (raw) apple cider vinegar. Ideally the vinegar should contain 'the mother', strands of proteins, enzymes and friendly bacteria that give it a murky, cobweb-like appearance and can make it look slightly congealed).

Or enjoy a salad made with a handful of bitter greens such as rocket, dandelion leaves, frisée or chicory and Apple Cider Dressing (see page 234), which will have a similar enzyme-stimulating effect.

Buckwheat Blinis

(makes 10-12 medium-sized blinis)

Despite its name, buckwheat is not related to wheat and is gluten-free. It is also a good source of minerals and antioxidants. These blinis look like a cross between a drop scone and a flat crumpet and make a brilliant savoury or sweet brunch.

90g buckwheat flour (or another wholegrain flour)
1 tsp baking powder
Generous pinch of salt
1 egg
150ml any milk (e.g. coconut, dairy, soya, almond)

Put the flour in a mixing bowl with the baking powder and salt. Make a well in the centre and add the egg and a little of the milk. Start beating the egg and milk with a fork, gradually incorporating the flour and adding more milk until you have a smooth batter. Leave it to stand for 20 minutes.

Lightly grease a large flat-based frying pan or a flat griddle and place it over a medium heat. Drop dessertspoonfuls of the batter onto the pan – you can make 3 or 4 at the same time but leave room for them to spread. Turn them over when holes appear in the surface and the top starts to firm up (about 2-3 minutes).

Flip them over again and cook them for another 1-2 minutes before removing them from the pan. They're delicious hot off the pan or you can store them in an airtight container and pop them in the toaster briefly before serving.

Tip: You could spread with full-fat cream cheese or a non-dairy equivalent, and top it with pickled fish or smoked mackerel or salmon, a squeeze of lemon and freshly ground black pepper.

Cashew Nut Cheese

A delicious spread to try on one of our breads.

340g raw cashews, soaked in water and drained
120ml water
4 tbsp nutritional yeast
2 tbsp lemon juice
2 garlic cloves
1 tbsp live (raw) apple cider vinegar
1 tbsp Dijon mustard
Sea salt and pepper to taste

Blitz all the ingredients in a blender until you have a thick, creamy paste.

Tips: use this 'cheese' as an alternative to mayonnaise on a chopped boiled egg, or as a dip with vegetables or gluten-free crackers.

Walnut & Red Pepper Dip

2 red peppers, halved and deseeded
3 tbsp extra-virgin olive oil
150g walnuts, roasted
Bunch of coriander or parsley
Salt and pepper to taste

Preheat the oven to 200°C. Place the peppers on a baking tray and drizzle the olive oil over them. Bake them at the top of the oven for 30 minutes, then chop them roughly and put them in a blender with the remaining ingredients. Add more walnuts if you want a thicker spread. Serve it with a selection of crudités such as carrot batons, celery, cauliflower florets or courgetti, or spread on Flaxseed, Chia & Red Pepper Crackers (see page 228) .

Green Seaweed Dip

Seaweed is a significant part of many diets around the world, particularly in Asia, prized for its high nutritional content – it is rich in protein, fibre and omega 3 – and wonderful flavours. We are excited about finding tasty ways to incorporate more of it in our recipes.

1 yellow pepper
1 medium courgette
400g tin butter beans (or chick peas), drained
1 garlic clove, crushed
2 tbsp olive oil
2 heaped tbsp pesto sauce
½ pack nori sushi sheets (about 15g), diced into approx 1cm squares

Place a wok or large pan with a lid over a high heat and

scorch the whole pepper and the courgette for about 10 minutes, turning them occasionally. When they are charred in a few places, turn off the heat and leave them to soften in the pan with the lid on for about 5 minutes. Alternatively, char the vegetables under a hot grill. Dice the courgette and the pepper, after removing the stalk, seeds and any patches of very blackened skin. Then tip them into a food processor, along with the beans, garlic and pesto. Pulse for a minute or so, leaving a bit of texture.

Mix the diced nori seaweed with the vegetables, keeping aside a small handful for garnishing. Blitz briefly. Pour the dip into a bowl, scatter the remaining seaweed on top and serve it with a selection of crudités such as carrot batons, mini corn on the cob, cauliflower and/or broccoli florets, or Flaxseed, Chia & Red Pepper Crackers (page 228).

Gut-soothing Bone Broth

The lining of your intestine can be damaged by all sorts of things: infections, some medications or certain proteins. It is composed of a single layer of cells, and if the tight junctions that hold these cells together open up, you are prone to a condition called 'leaky gut' or 'increased intestinal permeability'. Eating collagen may help. This is found in meat, gristle, skin and bones, and boiling down bones releases the collagen into the broth.

Bone broth is hugely popular with the health and fitness

crowd, even among those who don't have a leaky gut, because it also contains lots of important nutrients. It is also an ideal aid to recovery from illness.

- 1 medium organic chicken or 1½kg chicken parts (bones in)
- 1 onion, chopped
- 3 carrots, chopped
- 2 celery sticks, chopped
- 2-3 courgettes, sliced
- 2 garlic cloves
- 2 tbsp live (raw) apple cider vinegar
- 2 tbsp coconut oil
- 2-3 slices root ginger
- 1 tsp ground turmeric
- Sea salt and black peppercorns

Place all the ingredients in a large saucepan with just enough water to cover them. Bring the pan to the boil, then lower the heat, put the lid on and let it simmer, ideally for 6-8 hours, but for at least 2-3 hours, topping up the liquid with water as required.

Take the chicken out and place it on a platter to cool. Removc all the meat from the carcass – it can be used for a chicken salad (see page 235).

This broth will keep in the fridge for 3-4 days or can be frozen for up to 1 month. It can also be sipped as a warm drink. It is especially good with a squeeze of lemon and a little sea salt.

Creamy Cauliflower & Jerusalem Artichoke Soup

(serves 4)

Cauliflower is a good source of vitamin C and contains compounds that stimulate detoxification enzymes, while Jerusalem artichokes are full of gut-friendly inulin fibre.

2 tbsp olive oil
1 onion, finely chopped
3 garlic cloves, roughly chopped
½ tsp ground turmeric or 1 tsp fresh root, grated
1 large head of cauliflower, cut into florets
150g Jerusalem artichoke, scrubbed, peeled and sliced
400ml tin full-fat coconut milk
600ml vegetable stock
2 tsp tamari sauce
Freshly ground black pepper
Handful of fresh coriander leaves, chopped

In a large saucepan on a medium heat, gently fry the onion, garlic and turmeric in the oil for 5 minutes. Add the cauliflower florets and artichoke and stir everything well, then pour in the coconut milk, stock and tamari sauce. Bring it to the boil and then immediately lower the heat and simmer gently for 15 minutes or until the vegetables are tender. Season with pepper to taste. Process the soup in a blender until it's smooth. Serve it with chopped coriander and a slice of Flaxseed Bread (see page 210).

Happy Guts Chicken Soup

(serves 4)

1 white onion, chopped
2 tbsp olive oil
1 small garlic clove, chopped
2 large boneless chicken thighs, diced
700ml water or Gut-soothing Bone Broth (see page 216)
1 tsp root ginger, chopped
Juice of 1 lime
4 medium carrots, sliced
1 red and 1 yellow pepper, deseeded and chopped
1 head of broccoli, chopped
1 tsp fresh thyme, chopped (or ½ tsp dried)
Handful of fresh coriander and parsley, chopped
2 tsp chives, chopped

Gently sweat the onion and garlic with the chicken in the olive oil in a large pan until everything is golden. Then add the rest of the ingredients, apart from the fresh herbs, and simmer on a low heat for 60-90 minutes or until the vegetables are soft and the flavours have blended. Season generously with salt and freshly ground black pepper.

Add the fresh herbs a few minutes before serving.

Carrot & Turmeric Soup

(serves 4)

Turmeric, our anti-inflammatory friend, adds colour and flavour to this delicious carrot soup.

3 tbsp olive oil
1 white onion, chopped
2 tsp ground turmeric, or 2cm fresh root, grated
1 tbsp root ginger, chopped
2 garlic cloves, chopped
500g carrots, cut into 2cm chunks
400ml Gut-soothing Bone Broth (see page 216)
or bouillon (e.g. Marigold)
Juice of 1 lime

Place a large pan over a medium heat and sweat the onion in the olive oil with a large pinch of salt for 5 minutes. Add the turmeric, ginger and garlic and cook for 1-2 minutes, then add the carrots and stir in the bone broth or bouillon. Bring it to the boil, then reduce the heat and let it simmer, covered, for about 25 minutes. With a hand blender, blend the soup until there are no lumps. Add the lime juice and a little more water if needed. Season with sea salt and pepper to taste.

Green Pea Soup
(serves 4)

500g frozen petits pois
2 ripe avocados
A few fresh mint leaves, chopped
2 spring onions, chopped
1 tbsp tamari sauce (or soy sauce)

Cook the petits pois in boiling water for 3-4 minutes, drain them, then blitz them with the rest of the ingredients. Season to taste.

Classic Greek Salad
(serves 2)

This old favourite makes the ideal nutritious salad – plenty of antioxidants, fibre and vitamins (including C, A and B2) from the veg; calcium, protein and fats from the cheese to keep you satiated and help you absorb the vitamins; and last but not least, the lovely bugs to feed that biome…

½ medium cucumber, deseeded and cut into 1cm slices
2 large ripe tomatoes or 4 medium vine ripened, cut into wedges

½ small green pepper, deseeded and finely sliced (optional)
Handful of Kalamata olives
½ small red onion, finely sliced
80g good-quality feta cheese
Leaves from a sprig of fresh oregano, finely chopped (or a generous pinch of dried)
2-3 mint leaves, finely chopped
2 tbsp extra-virgin olive oil
1 tbsp red wine vinegar

Spread the cucumber and tomatoes around the sides of a wide bowl. Add the sliced green pepper, if using, followed by the olives and red onion. Place the chunk of feta on top and sprinkle the herbs over it. Drizzle generously with olive oil (the salad should be coated, but not drowning in it) and vinegar. Finish with a pinch of fine sea salt such as Maldon and freshly ground black pepper.

Tip: Greek salads are traditionally served dressed, not tossed, with the cheese in one piece on top.

Carrot & Beetroot Salad

(serves 4)

Not only do beetroots add great flavour and a juicy crunch to a dish, they are also one of the few 'superfoods' that really deserve their title. They are rich in nitrates and have even been shown to lower blood pressure.

- 2 tbsp freshly squeezed orange juice
- 2 tsp freshly squeezed lemon juice
- 4-5 tbsp olive oil
- 2 tbsp tamari sauce
- ¼ tsp sea salt
- 300g carrots, grated
- 300g beetroots, peeled and grated (it is best to wear plastic gloves for this)
- 150g pumpkin seeds, toasted
- 100g hazelnuts, toasted (or 125g feta, if using dairy)
- 1 tsp coriander seeds

Whisk together the orange and lemon juice, olive oil, tamari sauce and salt to make the dressing. Toss the carrots and beetroot together in a salad bowl.

Put a small pan on the hob and gently fry the pumpkin seeds and hazelnuts, turning them frequently until they're golden. Throw in the coriander seeds for the last minute or so. When they are cool, scatter them over the salad. Crumble the feta over if using.

Phyto Salad Bowl

This salad will help you achieve a good balance of flavour and nutrients, particularly those all-important phytonutrients. While at least three-quarters of your bowl should be plant-based, you should also ensure you get the required fats and proteins. We suggest you choose:

- *3-4 portions of coloured veg (one of which can be substituted with fruit), such as*:

  1 sliced carrot

  ½ sliced pepper (red, orange, yellow)

  ½ sliced courgette

  5 baby tomatoes

  4 steamed asparagus spears

  3-4 artichoke or palm hearts

  ½ cup of radishes, mange touts or mushrooms

  Fruit – (approx ½ cup) strawberries, unpeeled pear or apple, papaya, mango, grapes, pomegranate, blueberries or raspberries

- *1-2 cups of any of the following greens, such as*:

  Spinach, mixed leaves, rocket, kale, broccoli, chicory, cauliflower, bok choy, sprouts, Swiss chard, cabbage

- *1-2 portions of protein, such as*:

  2 hard-boiled eggs

  Meat: chicken, turkey, cold meat (about 80g)

  Oily fish: tuna, salmon, mackerel, sardines; or white

fish such as trout, cod, haddock (about 100g)
Dairy: hard cheese, halloumi, goat's cheese, feta (30-60g)
Plant protein: a generous handful of lentils, beans (reduce both of these if you suffer from IBS or bloating), nuts, seeds, tofu, tempeh, hummus

- *2-3 portions of health-boosting fats, such as*:

½ small avocado
6 olives
1-2 tbsp dressing made with extra-virgin olive oil, sesame, walnut or rapeseed oil (see page 232)
Toasted seeds or nuts: pumpkin, pine kernels, sunflower, hazelnuts, cashews

- *1-2 portions of pulses, squash or wholegrains (optional):*

50g cooked quinoa, brown rice, whole barley or wild red rice
1 slice wholegrain bread: millet, spelt or rye (or gluten-free as required)
50g cooked beans, lentils or chickpeas
100g roasted, diced pumpkin or butternut squash

- *And for extra flavour*:

Fermented vegetables, such as sauerkraut or kimchee
Pickled vegetables, such as cornichons, jalapeño peppers
Seaweed: nori cut into strips, kelp/dulse flakes
Fresh herbs: coriander, mint, parsley, basil, etc

Get those phytonutrients

Phytonutrients are the natural components of plants that keep them healthy, protecting them from disease and damage. Their antioxidant and anti-inflammatory properties also provide significant health benefits for us. Eating them helps to preserve the microbial balance in the gut, keeping it in good working order.

Phytonutrients in fruit and vegetables come in a range of colours – green, yellow-orange, red, blue-purple and white. The key is to eat a wide variety of colours, aiming for two or more of each per day.

Phyto Burst Lunchbox

(serves 1)

1. *Colours*:

½ cup red cabbage, finely sliced

1 small-medium carrot, grated or finely sliced

½ cup red pepper, deseeded and sliced

4 tinned artichoke hearts, drained (they taste even better scorched on a hot griddle)

2. *Greens*:

1 cup spinach leaves

4-5 broccoli florets, steamed

3. *Proteins*:

2 hard-boiled eggs, roughly chopped

2 tbsp mixed seeds, toasted

4. *Health-boosting fats*:

Dressing, e.g. Apple Cider Vinegar Dressing (see page 234)

½ avocado

5. *Optional pulses, wholegrains or squash*:

e.g. 100g pumpkin (or butternut squash), cubed and roasted with a drizzle of olive oil

And for extra flavour:

Handful of fresh coriander, chopped

Maldon sea salt and pepper to taste

In a large bowl, or a Tupperware box if you're taking it to work, mix the cooled roast pumpkin with the spinach, broccoli, sliced red cabbage, carrot, red pepper and artichoke hearts. Add the boiled eggs, then scatter the seeds, diced avocado and coriander leaves on top. Make your dressing and put it in a small screw-top jar if you're taking it to work, so you can dress your salad at lunchtime.

Flaxseed, Chia & Red Pepper Crackers

(makes about 40)

These tasty crackers go well with cheese or a dip, such as Green Seaweed Dip (page 215) or hummus (see page 230). The only trouble is they are very moreish…

- 1 medium red pepper
- 1 tsp olive oil
- 200g ground flaxseed
- 60g ground almonds
- 60g chia seeds
- 1 dessertspoon tomato purée
- Zest and juice of ½ unwaxed lemon
- 1 tbsp tamari sauce (or soy sauce)
- 1 tsp Marigold bouillon (or ½ organic vegetable stock cube)

Preheat the oven to 150°C. Lightly coat the red pepper in the olive oil and roast it for 20 minutes or until it's soft. Take it out of the oven, and when it has cooled, cut it in half, remove the seeds and dice it.

Meanwhile, blend the rest of the ingredients in a food processor with 50ml water (or put them in a bowl and use a hand blender).

Add the diced red pepper and pulse again. If the dough is too dry and crumbly, add ½-1 tbsp water, a little at a time.

Roll the dough as thin as you can without breaking it, to about 3mm thick. This is easiest done between 2 sheets of parchment paper greased with coconut oil or butter. Take off the top layer of paper and place the dough on a baking sheet.

Lightly score it into squares then bake it for 20-30 minutes. Check it frequently towards the end to catch it as it starts to turn golden (you don't want it to get too brown).

At this point, remove it from the oven, turn it over and bake it for a further 20 minutes or so to crisp up, checking regularly so it doesn't burn. Then break it into squares, allow these to cool and store them in a sealed container.

Tip: If you have one, use a silicone-based baking mat instead of the bottom layer of baking paper as silicone tends to be more non-stick. If the crackers need crisping up when you come to eat them, simply pop them in the oven at 120°C for 5 minutes.

Rainbow Hummus

Basic Hummus (serves about 4)

250g tinned chickpeas, drained
Juice of ½ -1 lemon
1 tbsp tahini
1 tsp sea salt
1 garlic clove, finely chopped
5 tbsp olive oil

Blend all the ingredients in a food processor or with a hand blender until you have a creamy paste.

Now, add some colour...

Green Hummus (serves about 6)

Blend an extra garlic clove and 150g cooked peas with the hummus.

Purple Hummus (serves about 6)

Blend 2 peeled, chopped roasted beetroots and a few mint leaves with the hummus.

Red Hummus (serves about 6)

1 large red pepper, halved and deseeded
Pinch of chilli flakes (optional)

Place the red pepper under a hot grill for 15-20 minutes or until the skin is slightly charred in places and the flesh has softened. Roughly chop it and blend it with the hummus, along with the chilli, if using.

Yellow Hummus (serves about 6)

1 extra garlic clove
2 tsp ground turmeric
1 tbsp olive or coconut oil
1 small cauliflower

Preheat the oven to 175°C. Remove the leaves from the cauliflower. Chop it into 2-3cm pieces and toss it in the oil and turmeric powder. Spread it out on a baking tray and bake it for 15-20 minutes or until it's soft and just turning golden, but not charred. Blend it with the hummus.

Dressings (1-2 tbsp per serving)

Be generous with the oil in your dressings, not least because they will make salad taste better. Oil increases the absorption of some nutrients while slowing the absorption of starchy carbohydrates and sugars, helping you feel full for longer.

Anti-Inflammatory Turmeric Dressing

4 tbsp extra-virgin olive oil
Juice and zest of 2 unwaxed lemons
¼ avocado
1 garlic clove, minced
1 tbsp grated fresh turmeric, or 1 tsp of ground turmeric
1 tsp honey
Pinch of Maldon sea salt to taste

Whizz all the ingredients together with a blender. Add more avocado for a thicker consistency.

Cashew Nut Dressing

1 tbsp Cashew Nut Cheese (see page 214)
2 tbsp live (raw) apple cider vinegar

Beat the ingredients together in a bowl.

Simple Oil & Lemon Dressing

200ml extra-virgin olive oil
50ml lemon juice
½ tsp honey or maple syrup
½ tsp cayenne pepper or ½ red chilli, chopped (optional)

Whisk all the ingredients together in a jar.

Lemony Vinaigrette

Juice of ½ lemon
Zest of 1 unwaxed lemon
1 tbsp live (raw) apple cider vinegar
1 tsp Dijon mustard
1 tbsp preserved lemon, finely chopped
5 tbsp extra-virgin olive oil

Whizz all the ingredients together, except for the oil, with a blender. Pour in the oil slowly to required consistency.

Green Vinaigrette

1 tsp capers, rinsed
Small handful of fresh parsley
Small handful of fresh basil
Small handful of fresh mint

1 tbsp fresh chives, chopped
1 garlic clove, chopped
1½ tbsp live (raw) apple cider vinegar
5 tbsp (75 ml) extra-virgin olive oil

Whizz all the ingredients together with a blender.

Apple Cider Vinegar Dressing

200ml extra-virgin olive oil
50ml live (raw) apple cider vinegar
1 tbsp lemon juice
½ tsp honey or maple syrup
1 garlic clove, finely chopped

Whisk all the ingredients together in a jar.

Hummus Dressing

2 tbsp hummus
2-4 tbsp live (raw) apple cider vinegar
2 tbsp nutritional yeast
Squeeze of lime
5 tbsp extra-virgin olive oil
Salt and ground black pepper

Whisk all the ingredients together in a jar.

Main dishes

Asian Coleslaw with Leftover Chicken

(serves 4)

This brightly coloured dish includes lots of gut-friendly ingredients. You can use the chicken left over from the Gut-soothing Bone Broth to make it (see page 216).

4 medium carrots
1 small white cabbage
½ small red cabbage
1 mango, cut into slices
Couple handfuls leftover cooked chicken

If cooking the chicken from scratch:
3 medium chicken breasts (with skin and bone)
1 medium carrot, halved crosswise
1 medium celery heart, halved crosswise
3 sprigs of flat-leaf parsley
2 sprigs of fresh thyme
2 fresh bay leaves
½ tsp whole black peppercorns
1 tsp coarse salt

For the dressing:

2 tbsp tamari sauce

1 tbsp honey

1 garlic clove

Thumb-size piece of root ginger, finely chopped

2 tbsp rice vinegar

2 tbsp sesame oil

2 tbsp olive oil

If cooking the chicken from scratch, put the chicken breasts in a saucepan so that they fit snuggly, and cover them with 2cm water. Add the vegetables and herbs. Bring to the boil, then turn down the heat and simmer gently for 20-25 minutes, skimming foam from the surface as the chicken poaches. Remove it from the pan and allow it to cool, then tear it into strips. Keep the stock to use elsewhere.

To assemble the salad, cut the carrots into fine strips with a vegetable peeler or grate them quite thickly into a large bowl. Discard the outer leaves and the cores from the cabbages and shred the leaves as finely as you can. Add them to the carrots, then mix in the cooked chicken and the mango slices. Make the dressing by whisking all the ingredients together in a bowl. Pour it over the salad and toss everything together. Serve it with fresh coriander leaves and a squeeze of lime.

Tip: Put the dressing in a small screw-top jar if you are taking it to work, so you can dress your salad at lunchtime.

Prawn and Seaweed in Tomato Sauce with Courgetti

(serves 4)

Welcome to the wonderful world of seaweed. Rich in omega 3 and fibre, nori is easy to use and great with prawns.

3 tbsp olive oil
1 medium onion, finely chopped
2 garlic cloves, chopped
60g salted anchovies in oil, chopped
1 tsp dried thyme (or fresh if available)
2 x 400g tins chopped tomatoes
300g frozen prawns, defrosted
10g nori sushi sheets, cut into 1cm squares
4 large courgettes, spiralised (if you don't have a spiraliser, you can buy ready-made courgetti)

Sweat the onion and garlic in 2 tbsp oil over a low heat. After 5-7 minutes, add the anchovies and thyme and stir, so they form a paste with the onions. Then add the chopped tomatoes and simmer gently for another 10 minutes. Stir in the prawns and cook them for a further 5 minutes.

Boil the kettle so that when the sauce is ready you can quickly steam the courgetti (it will only take 1-2 minutes, or 1 minute if you boil it). It should be 'al dente' and not soggy! Drain it well and drizzle the remaining olive oil over it. Mix the chopped seaweed into the sauce and serve it on top of the courgetti.

Veggie Burgers
(serves 4)

Legumes, such as chickpeas and butter beans, are an important part of the Mediterranean diet and here you can enjoy them in the form of a tasty veggie burger, bursting with fibre and goodness.

1 onion, finely chopped
2 tbsp olive oil
2 carrots, grated
1 tsp ground coriander
1 tsp curry powder
1 tsp cumin seeds
400g tin chickpeas or butter beans, drained
1 egg, whisked (or 3 tbsp coconut milk)
Handful of fresh coriander, chopped
Juice and zest of 1 lime
1 heaped tbsp mixed seeds (e.g. sunflower, sesame and pumpkin)
Wholegrain gluten-free flour (e.g. buckwheat) for dusting

Sweat the onion and carrots in 1 tbsp olive oil for about 5-7 minutes then stir in the ground coriander, curry powder and cumin seeds and continue to cook gently for a few more minutes. Using a food processor or a hand blender, briefly pulse the onion mixture with the chickpeas (or butter beans) and the egg (or coconut milk) so they are well combined but still have some texture.

Mix in the fresh coriander, lime juice, grated zest and seeds. Season the mixture generously with salt and pepper. Shape it into 8 medium-sized patties, dusting your hands with flour first so that they don't stick. If you have time, leave the patties in the fridge for about 20 minutes to firm up.

When you are ready to eat, heat the remaining olive oil in a pan and gently cook the burgers for 10-12 minutes, turning them occasionally, until they are golden brown.

These are delicious with a big salad of mixed leaves, coriander, cucumber slices and avocado served with Lemony Vinaigrette (see page 233).

Salmon & Tomato Burgers

(serves 4)

6 cherry tomatoes or sunblush tomatoes
½ red chilli, deseeded and chopped
4cm root ginger, chopped
Bunch of fresh coriander
2 tsp capers, rinsed
Juice and zest of 1 unwaxed lemon
4 organic salmon fillets, skinned
1 tbsp wholegrain gluten-free flour (e.g. buckwheat), plus extra for dusting
2 tbsp olive oil

Using the lowest setting on your hand blender or food processor so that you keep some texture, briefly blitz the tomatoes with the chilli, ginger, coriander, capers, lemon zest and half the lemon juice.

Dice the salmon and stir it into the mixture, then pulse again, but only very briefly so that the salmon is just flaked, not mush. Add 1 tbsp flour to bind the mixture. Shape it into 8 smallish patties, dusting your hands with flour first so that they don't stick. Place them on a plate, cover them with clingfilm and leave them in the fridge to firm up for 20 minutes.

Heat the olive oil in a large flat-based pan. Brown the patties on both sides, then turn the heat right down and fry them gently for 8-10 minutes or until they're cooked through.

You might serve them with a large salad or greens, and a couple of spoonfuls of cooked quinoa.

Kashmiri Chicken Curry

(serves 4)

The dark chicken thigh meat is higher in both vitamins A and D (essential for your gut health and immune function) than the white meat.

1 tbsp coconut oil

450g skinless, boned chicken thighs

1 large onion, sliced
2 garlic cloves, finely chopped
1 green chilli, thinly sliced
2cm root ginger, peeled and grated
1 tsp ground coriander, or 6-7 seed pods
1 tsp ground cardamom
1 tsp ground turmeric
100g flaked almonds
100ml chicken stock or bouillon
250ml full-fat Greek yoghurt
Bunch of fresh coriander, chopped

Heat the oil in a large saucepan and brown the chicken pieces. Transfer them to a plate and set them aside. Sweat the onion, garlic, chilli and ginger in the same pan for 5-7 minutes or until they start to soften. Add the spices and cook for a further 5 minutes, then add the almonds and the chicken. Pour in the stock and the yoghurt and stir well. Season with salt and pepper to taste.

Cover the pan with a lid and simmer for 40 minutes over a gentle heat, stirring occasionally and adding more water if needed. Alternatively, cook the curry in a casserole in an oven preheated to 150°C. Stir in the coriander before serving.

This curry goes well with brown basmati rice or Cauliflower 'Rice' (see page 256) and steamed greens.

Goan Fish Curry with Seaweed

(serves 4)

This aromatic curry is packed with health-boosting nutrients, including seaweed. Cod is rich in selenium, iodine and choline, while turmeric, garlic and ginger are known for their antimicrobial properties – Indians have been using them for centuries to reduce the risk of gut infections.

2 tbsp coconut oil
1 tsp mustard seeds
2 tsp ground coriander
1 tsp ground cumin
½ tsp turmeric
½ tsp garam masala
1 medium onion, roughly chopped
2cm root ginger, roughly chopped
6 garlic cloves, roughly chopped
1 mild or medium red chilli, deseeded and chopped
1 tbsp live (raw) apple cider vinegar
400ml tin full-fat coconut milk
1 large tomato, finely diced
500g firm white fish (e.g. cod, pollock, haddock or hake), cut into large chunks
Handful of fresh coriander
8-10g nori sushi sheets

For a vegetarian option:

250g butternut squash, cooked and cut into chunks

250g tinned chickpeas, drained

Heat the coconut oil in a large saucepan and fry the spices for 1-2 minutes. Blend the onion, ginger, garlic and chilli together with the apple cider vinegar to make a paste. Add the paste to the pan containing the spices and cook for 2 minutes, until you see the oil separate, then add the coconut milk and the tomato and bring it to the boil.

Add the fish and simmer gently until it is cooked through (about 10-12 minutes). If doing the veggie option, tip in the sweet potato and chickpeas and gently heat them through for a few minutes. Chop a 3cm strip off the pack of nori seaweed, then cut the strip into ½cm-wide pieces and stir them into the curry.

For a fuller seaweed taste, add another, larger strip, 2-3cm wide. Check the seasoning – it may be salty enough with the seaweed.

Finally, stir through the fresh coriander, saving a few leaves to garnish.

Serve with Cauliflower 'Rice' (see page 255) or a couple of spoonfuls of brown basmati rice and a green veg.

Chicken Goujons
(serves 4)

You will be glad to hear that these taste nothing like the sad little nuggets you buy in takeaways. The browner meat of the chicken thighs is juicier and more flavoursome than the breast. It also contains more nutrients.

4 chicken thighs, boned and diced
1 tsp dried oregano
1 tsp garlic powder
1 tsp onion powder
½ tsp sea salt
100g ground almonds
1 egg, beaten
2 tbsp olive oil (or coconut oil)

Cut the chicken into 2cm strips. Put the rest of the ingredients apart from the oil and the egg into a bowl and mix them well together. Dip the chicken pieces in the egg, then coat them evenly with the spice mixture.

Heat the oil in a large pan and fry the chicken pieces, turning them occasionally until they're golden and cooked through. Serve them with some homemade hummus, sliced avocado and a large green salad.

Harissa & Beetroot Salmon
(serves 2)

Beetroot and salmon work surprisingly well together. You will find za'atar and harissa paste in most big supermarkets.

2 organic salmon fillets
1 tbsp za'atar
200-250g beetroot
2 medium sweet potatoes
1 fennel bulb
1 tsp ground cumin
4 tbsp olive oil
1 carrot
Juice of 1 lime
2 tbsp harissa paste
1 tbsp flaked or chopped almonds, toasted

Preheat the oven to 170°C. Rub the salmon with the za'atar, a pinch of salt and a grind of pepper and put it aside.

Peel the beetroot, halve or quarter it if it's large, and cut it into ½cm slices. Peel the sweet potato, quarter it lengthwise, and dice it into 2cm chunks. Trim and quarter the fennel bulb, and put it in a medium to large baking dish, along with the beetroot and sweet potato. Scatter the cumin over it, season it with salt and pepper and drizzle 1 tbsp olive oil over the surface. Place the dish on the middle rack of the oven for around 25 minutes, or until everything is soft, stirring once or twice.

Meanwhile, grate the carrot into a bowl and mix it with half the lime juice. In another bowl, whisk together 2 tbsp olive oil, the harissa, the rest of the lime juice and a pinch of salt. When the vegetables are ready, remove them from the oven, pour the dressing over them, add the grated carrot and toss everything together. Place a frying pan over a medium heat and fry the salmon fillets in the remaining oil for 3 minutes, skin side down. Then flip them over and cook them on the other 3 sides for a minute each.

Serve the salmon on top of the baked vegetables, with the almonds scattered over it.

Mackerel or Cod in Salsa Verde

(serves 2)

Oily mackerel is the perfect foil for this feisty-flavoured sauce – and a favourite of Michael's. It's also a great source of omega 3 and vitamins B12 and D.

2 mackerel or cod fillets
Green Vinaigrette (see page 233)

Preheat the oven to 180°C. Place the fish in a baking dish lined with greaseproof paper, rub a good tablespoonful of the vinaigrette over it and bake it for 20 minutes.

Serve it with more of the vinaigrette, some roasted veggies and seasonal greens.

Turkey Burgers
(serves 4)

Turkey is a high-quality protein, which helps you feel full for longer. Frozen spinach is just as nutritious as fresh.

1 onion, chopped
2 garlic cloves, crushed
2 tsp dried oregano
8 large basil leaves
3 handfuls of baby spinach (or 2 balls of frozen spinach, defrosted)
Zest of 1 lemon
½ tsp sea salt
Freshly ground black pepper
3 tbsp extra-virgin olive oil
400g turkey mince

Blitz the onion, garlic, oregano, basil, spinach, lemon zest, salt and pepper in a food processor with 1 tbsp olive oil, just briefly to retain some texture. Place the mixture in a large bowl and combine it with the turkey mince, using either your hands or a rubber spatula. Shape the mixture into 8 patties and pop them in the fridge for at least 20 minutes to firm up. Place a griddle or a frying pan over a medium-high heat and fry the patties in the remaining oil until they've browned – they'll need approximately 5 minutes on each side. Serve them with a large salad, some hummus and griddled courgettes.

Beef Goulash

(serves 4)

Like all red meats, beef is loaded with iron, which many people, particularly women, are chronically deficient in. It's a high-quality protein and contains all 8 essential amino acids. Do buy good-quality grass-fed if you can.

- 4 tbsp olive oil
- 1 large white onion, chopped
- 2 large carrots, cut into batons
- 1 large green pepper, deseeded and sliced
- 3 garlic cloves, crushed
- 500g braising steak, diced
- 1 tbsp paprika
- 3 tbsp tomato purée
- 3 bay leaves
- 400ml Gut-soothing Bone Broth (see page 216) or organic beef stock
- 400g tin chopped tomatoes
- 1 tbsp live (raw) apple cider vinegar

Preheat the oven to 160°C. Heat 2 tbsp oil in a large casserole with a well-fitting lid and fry the onions gently for 5-7 minutes. Add the carrots, green pepper and garlic and cook for few more minutes.

Place 2 tbsp oil in another pan over a high heat and brown the meat on all sides. Season it and add it to the vegetables in the casserole, along with the paprika, tomato purée and bay leaves. Pour the stock or broth into the pan you used for the meat and stir for a minute or so, scraping the bottom, to incorporate all the juices from the meat, then add it to the casserole, along with the tomatoes and vinegar. Bring the goulash to a simmer, then cover it and place it in the middle of the oven for 2½ -3 hours, taking it out occasionally to give it a stir, and adding more water if it is drying out.

Serve it by itself with a dollop of full-fat organic Greek yoghurt or a non-dairy equivalent, or with a generous serving of green vegetables and a few reheated new potatoes (as these contain some resistant starch).

Caribbean Coconut & Vegetable Curry

(serves 6)

A colourful and filling vegetarian dish bursting with flavour. It looks like a lot of ingredients, but they are easy to put together and you just stick it in the oven. Adding black pepper significantly enhances the benefits of turmeric.

4 tbsp coconut oil

1 large onion, diced

1 tsp each of ground coriander, cumin, turmeric

½ tsp ground cinnamon or 1 cinnamon stick

2 garlic cloves, chopped

4-5cm root ginger, finely chopped
1 chilli, deseeded and finely chopped
400ml tin full-fat coconut milk
150ml vegetable stock
2 limes, 1 for juice and 1 cut in wedges
4-5 large pitted dates, chopped (or 2 tsp maple syrup)
400g butternut squash, deseeded and diced into 2-3cm pieces
2 red peppers, deseeded and cut into strips
400g tin black eye beans, drained and rinsed (optional)
250g cherry tomatoes
200g kale, chopped and thick stalks removed
Bunch of fresh coriander, chopped
120g cashew nuts

Preheat the oven to 180°C. In a large casserole with a tight-fitting lid, heat the coconut oil and fry the onion on a moderate heat for 5-7 minutes or until it's golden. Then add the coriander, cumin, turmeric and cinnamon, followed by the garlic, ginger and chilli, and fry them for 2-3 minutes. Next pour in the coconut milk, stock and lime juice, along with the dates, butternut squash, red peppers, beans and tomatoes. Cover the casserole and put it in the oven.

After about 20-25 minutes, remove the casserole from the oven and stir in the kale. Return it to the oven for a few more minutes until the kale is cooked. Season with salt and plenty of freshly ground black pepper. Stir in some of the coriander and scatter the rest on top along with a sprinkling of cashews, and serve with the lime wedges.

Tip: For extra flavour, use whole spice seeds and grind them in a pestle and mortar yourself before cooking. For convenience, you can buy pre-prepared, diced butternut squash. Skip the beans if you're avoiding pulses.

Lamb & Sweet Potato Stew

(serves 5-6)

Lamb is a succulent outdoor-reared meat, and makes a great stew. Slow-cooked food tends to be easier to digest.

800g stewing lamb, diced
½ tsp each of cinnamon, nutmeg, ground cloves and ginger (or 2 tsp mixed spice)
3 tbsp olive oil
2 large red onions, sliced
4 rashers of bacon, cut into strips
3 bay leaves
500ml Gut-soothing Bone Broth (see page 216) or organic beef stock
6 celery sticks, diced into 1-2cm pieces
4 medium sweet potatoes, diced

Preheat the oven to 160°C. Place the diced lamb in a bowl and toss it with the spices and some seasoning. Leave it to marinate overnight or for at least a couple of hours.

Put the olive oil in a large casserole with a tight-fitting lid and sauté the onions for 5-7 minutes or until they start to turn golden. In a non-stick saucepan, sear the meat over a high heat, then tip it into the casserole dish along with any juices.

Add the bacon, bay leaves and broth or stock. Bring it to a simmer, then place it in the oven for 2-3 hours, checking occasionally and adding more water if it starts to dry out.

Halfway through the cooking time, add the celery and sweet potato. Serve the stew with cooked quinoa and an enzyme-stimulating green salad (see page 212).

Creamy Cashew Mushrooms

(serves 2)

Mushrooms are an underrated food source. They are high in fibre, rich in B vitamins, selenium and copper, and also very low in calories. Leaving them out in the sun for an hour or two will significantly boost their vitamin D levels.

1 tbsp coconut oil or butter

1 large onion, diced

1 large garlic clove, crushed
250g chestnut or mixed exotic mushrooms, sliced
½ tsp each of chilli flakes, dried thyme, crushed
  pink peppercorns, paprika
30-40g cashew nuts
½ mushroom stock cube (or 1 vegetable stock cube)
160ml tin coconut cream
Juice of ½ lemon
Small bunch of fresh parsley, chopped

In a medium-sized pan, gently sweat the onion in the coconut oil for 5-7 minutes, then add the garlic, mushrooms, chilli flakes, thyme, pink peppercorns and paprika.

Add the cashew nuts and cook on a medium heat for a few more minutes. Crumble in the stock cube, then stir in the coconut cream and the lemon juice. Season with salt and a generous amount of freshly ground black pepper. Simmer for a couple of minutes until everything melds together into a delicious, aromatic stew.

Serve with sliced spring greens or courgetti. If you're hungry, add 1-2 tbsp cooked brown basmati rice per person. Ideally the rice should be cooked, cooled for 12 hours then reheated to boost resistant starch.

Baked Rainbow Ratatouille

(serves 4)

This rainbow-coloured ratatouille is bursting with a whole range of phytonutrients.

4 beetroots

4 sweet potatoes

3 red onions

2 courgettes

1 fennel bulb

1 small pumpkin (or ½ butternut squash), deseeded and quartered

2 yellow peppers

250ml tomato sauce (see below)

1 tsp Maldon sea salt

Freshly ground black pepper

2 tbsp olive oil

For the tomato sauce:

2 tbsp olive oil

1 medium onion, sliced

2 garlic cloves, chopped

Bunch of fresh basil, leaves only, chopped

1 tbsp fresh oregano leaves, chopped (or 1 tsp dried)

6 medium-sized fresh tomatoes, chopped

400g tin chopped organic tomatoes

1 tbsp balsamic vinegar

To make the tomato sauce, place a saucepan over a medium heat, add the oil and fry the onion gently for 6-7 minutes until it's soft and golden. Stir in the garlic, basil and oregano, followed by the fresh and tinned tomatoes and the vinegar. Season the sauce with a pinch of salt and pepper and let it simmer for 15-20 minutes. Blend it briefly and set it aside (it will keep in the fridge for about 5 days).

Preheat the oven to 190°C. Liberally cover the bottom of a large baking dish with your homemade tomato sauce (you only need about half of it for this recipe). Cut all the vegetables into 1cm slices, except the courgette and peppers, which need to be thicker (more like 2cm). Starting from the outer edge of the dish, arrange the vegetables by colour: slices of beetroot, then sweet potato, courgette and so on until you fill it. Season with salt and pepper and drizzle the olive oil over the surface. Bake it for 30-40 minutes.

This works well with some steamed kale or a large green salad and avocado, dressed with Green Vinaigrette (see page 234).

Cauliflower 'Rice'

(serves 2)

This is a low-carb alternative to rice, with far more nutrients. It's a great way of getting more cauliflower into your diet.

½ cauliflower
30g pine nuts
½ tsp sea salt
Ground black pepper
Juice of ½ lemon
2 tsp tahini
½ tsp cumin seeds
½ tsp ground turmeric
½ tbsp olive oil

Preheat the oven to 200°C. Either grate the cauliflower, or cut it into florets and blitz it in a food processor so that it resembles grains. Toss it in a bowl with the rest of the ingredients, then spread it out in a thin, even layer on a baking tray. Roast it in the oven for 10-12 minutes, turning it halfway through – this dries the 'rice' out, giving it a fluffy texture.

Purple Sauerkraut
(makes enough for one 500ml jar)

Sauerkraut is one of the easier fermented foods to make and is packed full of probiotic bacteria.

200g beetroot, grated (use carrot if you don't like beetroot)
300g red cabbage, very finely sliced
½ small apple, peeled, cored and finely diced
½ tsp caraway seeds
1 tsp fennel seeds
1 tsp coriander seeds
2 tsp sea salt

Wear rubber gloves to avoid staining your hands red. Put the beetroot or carrot in a good-sized bowl along with the cabbage and apple. Add the seeds and salt and massage them well into the vegetables until they start to exude water; then firmly crush the vegetables with your hands or a blunt object like a pestle or the end of a rolling pin.

Pack the mixture into a sterilised resealable 500ml glass jar, leaving about 2cm space at the top to allow for the mixture to bubble and fizz – remember it's alive!

You need to push the vegetables down hard so that they are totally submerged in their juices. You could insert a weight to hold them down (a boiled beach pebble works well, or a chunk of cabbage stalk).

Leave the sauerkraut at room temperature for up to 3

weeks, but for at least 3 days. The longer you leave it to ferment, the more sour and distinctive its flavour.

Check and taste it regularly. Open the jar daily for the first 2-3 days to release the carbon dioxide, and keep checking that the vegetables are fully submerged in the brine. Continue to check your ferment regularly and push the contents below the surface if needed. If you find it's drying out, add ½ tsp salt to 100ml filtered water, and use this to top it up.

Don't be alarmed if yeasts form on the surface of your pickle, you can scrape these off. When you are ready to eat it, remove any discoloured vegetables from the top. You can keep it in the fridge for several months.

Tip: It is important to use filtered water because most of the chlorine will have been removed – chlorine kills the bacteria needed for fermentation.

Treats

Chocolate Aubergine Brownies

(makes 12 small squares)

Strange as it may sound, the aubergine works brilliantly with the dark, flavonoid-rich chocolate. Truly yummy.

1 medium aubergine (200g), peeled and diced
150g dark chocolate (minimum 70 per cent cocoa solids), broken into pieces
60g coconut oil
60g soft pitted dates, diced
½ tsp salt
3 eggs, beaten
1 tsp baking powder
80g ground almonds

Preheat the oven to 170°C. Steam the aubergine for 15 minutes until it's soft (or microwave it in less time). Put it in a medium-sized mixing bowl and stir in the chocolate and coconut oil. The warm aubergine will melt the chocolate and oil. Add the chopped dates and salt. Using a hand blender or a food processor, blitz the mixture until it's smooth. By now it should be cool enough to add the eggs and baking powder. Blitz again for another minute or so, then mix in the ground almonds. Spread the mixture onto a medium-sized baking tray lined with greaseproof paper and bake in the oven for about 20 minutes. It is cooked when a knife comes out clean.

Serve the brownies with Raspberry Chia Jam (see page 209) and/or full-fat live organic Greek yoghurt.

Tip: Alternatively, to make cupcakes, divide the mixture into a 12-hole cupcake tray lined with paper cases, and bake for about 15-20 minutes.

Quick Baked Apple Slices

(serves 2)

A quick, nutritious go-to dessert which takes a few minutes to prepare – it goes in the oven just before you start your meal and is ready in 20 minutes.

1 tbsp coconut oil, melted, plus extra for greasing
3 large sweet eating apples
1 heaped tsp ground cinnamon
2 tbsp pistachio nuts

Preheat the oven to 170°C. Lightly grease the base of a baking dish with coconut oil. Core the apples, cut them into quarters and cut each quarter into 2-3 wedges, leaving the skin on. Place the wedges in a slightly overlapping layer in the baking dish, drizzle the coconut oil over them and sprinkle cinnamon on top.

Bake the apples in the oven for 10 minutes, then scatter the pistachio nuts over them and bake them for a further 5-10 minutes. They are ready when they start to brown around the edges. Serve them as they are, or with a generous dollop of full-fat live organic Greek yoghurt.

PHASE 1 – REMOVE & REPAIR MEAL PLANNER

This phase is gluten and dairy-free, very low in grains and pulses (except hummus) and a maximum of two servings of fruit a day. Plants are centre stage – think variety and colour. Avoid snacking between meals and try to fast overnight for a good 12 hours.

	Breakfast
SUN	2 eggs scrambled with wilted spinach and tomatoes. Slice of *Green Flaxseed Bread* spread with coconut oil.
MON	*Leaky Gut Healing Smoothie*.
TUE	*Nutty Cinnamon Granola* with *Nut Milk** and a handful of berries.
WED	*Pumpkin Porridge* and handful blueberries.
THU	2 slices *Breakfast Bread* with *Rainbow Hummus*, smoked salmon & chilli flakes.
FRI	*Chia Pot* with 1 tbsp each of blueberries and ground almonds.
SAT	*Healthy Gut Green Smoothie* with 1 slice *Breakfast Bread* and 1 tbsp of almond or cashew nut butter.

Lunch	Dinner
Enzyme-stimulating Green Salad, followed by *Carrot & Turmeric Soup* with *Flaxseed, Chia & Red Pepper Crackers* with rainbow hummus.	*Turkey Burgers* with sweet potato wedges, green leaf salad and *Anti-inflammatory Turmeric Dressing*.
Left-over *Turkey Burgers* with *Carrot & Beetroot Salad*.	*Creamy Cashew Mushrooms* with steamed broccoli and kale with chilli flakes.
Enzyme-stimulating Green Salad followed by *Happy Guts Chicken Soup*.	*Salmon & Tomato Burgers* with broccoli and *Hummus Dressing*.
Phyto Salad Bowl with *Flaxseed, Chia & Red Pepper Crackers*.	*Kashmiri Chicken Curry* with *Cauliflower 'Rice'*.
Enzyme-stimulating Green Salad followed by *Prawn & Seaweed in Tomato Sauce with Courgetti*.	*Baked Rainbow Ratatouille* with green leaf salad and *Lemony Vinaigrette*.
Enzyme-stimulating Green Salad followed by left-over *Rainbow Ratatouille* with *Rainbow Hummus* and 1 slice *Green Flaxseed Bread*.	*Cod in Salsa Verde* with steamed green beans and baby new potatoes.
Phyto Salad Bowl with *Flaxseed, Chia & Red Pepper Crackers*.	*Enzyme-stimulating Green Salad* followed by *Beef Goulash* with wilted greens or kale.

* Nut Milk – either homemade as per recipe (see page 204) or shop-bought coconut (carton not can), almond, hemp or hazelnut

PHASE 2 – REINTRODUCTION MEAL PLANNER

Include more prebiotic plants: asparagus, banana, Jerusalem artichokes, leeks, onions, bok choy and pulses. Add probiotic foods, including fermented foods: sauerkraut, yoghurt and kefir. Add a few more grains, but still do your best to avoid wheat – go for rye, spelt or a good sourdough that has been fermented.

	Breakfast
SUN	*Chia Pot* with 1 tbsp raspberries and small glass of *Kefir*.
MON	*Nutty Cinnamon Granola* with *Homemade Yoghurt* and handful blueberries.
TUE	2 poached or scrambled eggs, handful washed spinach leaves and 6 baby tomatoes. A slice of toasted rye bread and butter.
WED	*Creamy Cashew Nut & Banana Breakfast Pot*. Slice *Breakfast Bread* with *Raspberry Chia Jam.*
THU	*Kiwi and Chia Seed Smoothie*. Slice *Breakfast Bread* with nut butter and handful blueberries.
FRI	*Homemade Yoghurt* with 2 tablespoons of ground flaxseed, tablespoon of blueberries and half a banana.
SAT	2 boiled eggs with 4-6 steamed asparagus spears and slice of toasted *Green Flaxseed Bread* or sourdough.

Lunch	Dinner
Classic Greek Salad and *Flaxseed, Chia & Red Pepper Crackers*.	Roast chicken and variety of roasted vegetables. *Quick Baked Apple Slices*.
Asian Coleslaw with Leftover Chicken (using previous night's roast).	*Harissa & Beetroot Salmon* with steamed asparagus and broccoli drizzled in olive oil.
Enzyme-stimulating Salad followed by *Cauliflower & Jerusalem Artichoke Soup* with toasted *Green Flaxseed Bread*, half an avocado and drizzled olive oil.	*Turkey Burgers* with multi-coloured salad, *Lemony Vinaigrette* and a tbsp *Purple Sauerkraut.*
Phyto Burst Lunchbox with some *Purple Sauerkraut*.	*Goan Fish Curry with Seaweed*, brown basmati rice.
Enzyme-stimulating Salad followed by 2 slices toasted *Green Flaxseed Bread* with tinned sardines, tomato and watercress.	*Veggie Burgers* with sweet potato wedges and 3-4-colour salad with *Anti-inflammatory Turmeric Dressing*.
Enzyme-stimulating Green Salad followed by left-over *Veggie Burgers* with crudités* and *Rainbow Hummus*.	*Chicken Goujons* with *Walnut & Red Pepper Spread*, brown rice and *Carrot & Beetroot Salad*.
Phyto Salad Bowl.	*Lamb & Sweet Potato Stew* with steamed greens or kale.

*crudités = chopped up sticks of cucumber, peppers, carrots, asparagus, courgette, little gem lettuce, radishes, steamed broccoli or cauliflower florets

CLEVER GUTS DAILY FOOD & SYMPTOMS DIARY

Meal	Time	Food & Drink – content & quantity
Breakfast		
Snack?		
Lunch		
Snack?		
Supper		
Snack?		

Date: ....

Symptoms: e.g. bloating, abdominal pain, nausea, sickness, diarrhoea, brain fog, irritability, headache	**Symptom time & duration**	**Other factors:** including stress, poor sleep, exercise, illness, medicines & remedies

Foods high in resistant starch

Resistant starch is a form of starch that is resistant to being broken down in your small intestine. Instead of causing your blood sugars to spike, it acts more like fibre, reaching the colon largely intact. It is a valuable source of nutrients for the 'good' bacteria in your gut. A diet rich in resistant starch has been linked to reduced gut inflammation and reduced risk of type 2 diabetes and obesity.

Although it acts in a similar way to fibre, it is not always found in the same foods. Breakfast cereals like All Bran, which have decent levels of fibre, do not contain any resistant starch.

A word of warning: if you are currently on a typical Western diet, low in fibre, you should not suddenly hit your gut with a massive dose of resistant starch. Otherwise you could find yourself becoming uncomfortably bloated and windy. Ease into it. You can afford to go up to around 35g of resistant starch a day, but I wouldn't go much higher. Our menus contain foods with moderate levels of resistant starch.

Below is a list of foods with the highest to the lowest levels of resistant starch, given as grams per 100g of food. You will see from this list that there is a striking difference between unripe and ripe bananas.

You will also notice something that appears to be very confusing about resistant starch: while some foods become *more* resistant when cooked and cooled, others become less so. Potatoes and rice, for example, more than double their levels of resistant starch when cooked and cooled. Oats, on the other hand, have very high levels when raw (11g) but these fall to almost nothing (0.3g) when cooked. I occasionally have raw oats with yoghurt and berries for breakfast.

Potato starch behaves quite differently from fresh potatoes in that it has far higher levels of resistant starch when raw. Levels drop to almost zero when it is cooked. You buy it from health food stores or over the internet as a white powder. It is totally bland and incredibly cheap.

Potato starch is also the odd one out on this list because it is highly processed. I wouldn't normally recommend processed foods but I am interested in the potential link between potato starch and insomnia. I have started taking a teaspoon mixed with kefir, yoghurt or half a glass of milk before bed.

You can find a longer list of foods and their resistant starch levels at http://freetheanimal.com/wp-content/uploads/2013/08/Resistant-Starch-in-Foods.pdf

The following are averages, in grams of resistant starch per 100 grams of food, based on different studies. Because the research is new there are no figures for what happens to rice or potatoes that are cooked, cooled, then reheated.

High

Potato Starch	72	Cashew nuts	13
Unripe bananas	19	Oats, rolled, uncooked	11

Medium

Peas – cooked	6.7	Hummus	4.1
Lentils - cooked	6.6	Baked Beans	3.6
Potatoes – cooked & cooled	5.8	Plantain, Cooked	3.5
Brown rice – cooked & cooled	5.5	Ripe Banana	3.2
Chestnuts	4.9	Kidney beans, canned	2.1
Rye bread	4.3	Sourdough bread	2.1
Peanuts	4.2		

Low

Puffed rice crackers	0.8	Baked potatoes	0.6
Baked potatoes	0.6	Croissant	0.4

Which probiotics work and for what

The following information is taken from Medline Plus, part of the US National Library of Medicine. It is a reliable source of the latest information.

Are probiotics safe?

In healthy people, yes. But you are advised not to take them if you are pregnant, have recently had surgery or have a faulty immune system.

The two most studied strains are *Lactobacillus* and *Bifidobacteria*, which are found in fermented foods like yoghurt, kefir and cheese. But it is more complicated than that because there are so many different species of each. Just because one species of *Lactobacillus* might help prevent diarrhoea, it doesn't mean that another will.

How effective are they?

The Natural Medicines Comprehensive Database rates treatments as: Effective, Likely Effective, Possibly Effective, Possibly Ineffective, Likely Ineffective, Ineffective and Insufficient Evidence to Rate. None of the probiotics you can buy in capsule form have sufficient studies behind them to fall into the 'Effective' category, while a number fall into the 'Likely Ineffective' or 'Ineffective' category.

The following list only includes those considered 'Likely' or 'Possibly Effective'.

Bifidobacteria

Constipation. *Bifidobacterium longum*, whether in the form of capsules or from fermented foods, can increase bowel movements in constipated adults. This strain also seems to help prevent the flu in elderly people.

Irritable bowel syndrome (IBS). Taking VSL#3, a product containing *Bifidobacterium*, *Lactobacillus*, and *Streptococcus* may decrease bloating in people with IBS. It is the probiotic that my experts favoured.

Lung infections. Taking HOWARU Protect, which contains *Lactobacillus acidophilus* and *Bifidobacterium*, may help reduce and shorten symptoms of fever, cough and runny nose.

Diarrhoea in children caused by rotavirus. Taking *Bifidobacterium bifidum* seems to help prevent rotaviral diarrhoea when used with other bacteria such as *Streptococcus thermophilus* or *Bifidobacterium* Bb12.

Traveller's diarrhoea. Taking *Bifidobacterium* may help prevent traveller's diarrhoea when used with other bacteria such as *Lactobacillus acidophilus* or *Lactobacillus bulgaricus.*

Lactobacillus

Diarrhoea in children caused by rotavirus. This is most likely to be effective if you use at least 10 billion colony-forming units (CFUs – see page 273) for the first 48 hours after diarrhoea starts.

Hay fever. Taking at least two billion CFUs of *Lactobacillus paracasei* daily for 5 weeks may improve quality of life for people with grass pollen allergy.

Preventing diarrhoea caused by antibiotics. *Lactobacillus* seems to reduce the risk of diarrhoea caused by antibiotics. Giving children *Lactobacillus* GG (Culturelle) along with antibiotics is a good idea.

Eczema (atopic dermatitis). *Lactobacillus* GG may reduce symptoms of eczema in infants who are allergic to cow's milk. *Lactobacillus sakei, Lactobacillus plantarum*, and a combination of freeze-dried *Lactobacillus rhamnosus* and *Lactobacillus reuteri* also seem to reduce eczema symptoms in children under 13. But *Lactobacillus paracasei* doesn't seem to help eczema.

Preventing diarrhoea due to cancer treatment (chemotherapy). There is some evidence that patients with cancer of the colon or rectum have less severe diarrhoea and have shorter hospital stays if they take *Lactobacillus rhamnosus* and *Lactobacillus* GG (Culturelle).

Irritable bowel syndrome (IBS). *Lactobacillus acidophilus* can improve symptoms of IBS such as bloating and stomach pain.

Lung infections. Children under 6 who attend day-care centres seem to get fewer and less severe lung infections when given milk containing *Lactobacillus* GG or a specific combination product containing both *Lactobacillus acidophilus* and *Bifidobacterium* (HOWARU Protect).

Rheumatoid arthritis (RA). Research shows that taking *Lactobacillus casei* for 8 weeks reduces tender and swollen joints in women with RA.

What to look out for when buying a probiotic:

A high colony-forming units (CFU) count

Your microbiome contains around 40 trillion microbes, so to make an impact you are going to need a probiotic containing lots of bacteria. In other words, a high CFU count. Ideally you should be looking for a product that contains at least 20 billion CFUs and preferably more. Many commercial products are not even close to this minimum standard. Fermented foods are likely to hugely exceed these figures, though it is hard to generalise.

Multiple Strains

Just as you want diversity in your gut, so you want a diversity of strains of microbes in your probiotic. The joy of fermented foods is that you are guaranteed diversity. If you are buying a probiotic you should be looking for one with at least 6 different strains.

Substrains

Lactobacillus acidophilus is not the same as *Lactobacillus casei*. If the probiotic you are thinking of buying doesn't list substrains then it may well be ineffective.

- On our website cleverguts.com we list some of the better products

For further resources, go to cleverguts.com

Endnotes

1 Revised estimates for the number of human and bacteria cells in the body. Sender, Fuchs, Milo et al. *PLoS Biol*, 2016 (http://journals.plos.org/plosbiology/article?id=10.1371/journal.pbio.1002533)

2 Weight loss with a low-carbohydrate, Mediterranean or low-fat diet. Shai I, et al, *New England J of Med*, 2008 (http://www.nejm.org/doi/full/10.1056/NEJMoa0708681#t=article)

3 Successful weight loss maintenance includes long-term increased meal responses to GLP-1 and PYY 3-36. Lepsen, Lungren, Holst et al, *Eur J Endocrinol*, 2016 (https://www.ncbi.nlm.nih.gov/pubmed/26976129)

4 Surface area of the digestive tract revisited. Helander HF, Fandriks L, *Scandinavian J Gastroenterology*, 2014. (https://www.ncbi.nlm.nih.gov/pubmed/24694282)

5 Non-celiac gluten sensitivity among patients perceiving gluten-related symptoms. Capannolo A, Viscido A, Barkad MA, et al, *Digestion*, 2015

6 A purified membrane protein from Akkermansia muciniphila or the pasteurized bacterium improves metabolism in obese and diabetic mice. *Nature Medicine*, 2016

7 http://www.mayoclinic.org/digestive-system/expert-answers/faq-20058340

8 Association between caesarean birth and risk of obesity in offspring in childhood, adolescence, and early adulthood. *JAMA Pediatrics*, 2016

9 Environmental spread of microbes impacts the development of metabolic phenotypes in mice transplanted with microbial communities from humans. *ISME Journal*, 2016

10 Personalised nutrition by prediction of glycaemic response. Zeevi et al. *Cell*, 2015

11 Is eating behavior manipulated by the gastrointestinal microbiota? Evolutionary pressures and potential mechanisms. *BioEssays*, 2014

12 Ingestion of Lactobacillus strain regulates emotional behavior and central GABA receptor expression in a mouse via the vagus nerve. Bravo et al, *PNAS*, 2011

13 Human metabolic phenotypes link directly to specific dietary preferences in healthy individuals. Rezzi et al, *J Proteome Res*, 2007

14 Clinical and metabolic response to probiotic administration in patients with

major depressive disorder. Akkasheh et al, *Nutrition*, 2016 Mar; 32(3):315-20. doi: 10.1016/j.nut. 2015.09.003

15 The spread of obesity in a large social network over 32 Years. Christakis and Fowler, *New England J of Med*, 2007 (http://www.nejm.org/doi/full/10.1056/NEJMsa066082#t=article)

16 Helminth therapy for induction of remission in inflammatory bowel disease. Garg et al, *Cochrane Database Syst. Rev*, 2014. (https://www.ncbi.nlm.nih.gov/pubmed/24442917)

17 Primary prevention of cardiovascular disease with a Mediterranean diet. Estruch et al. *New England J of Med*, 2013

18 Understanding the impact of Omega-3 rich diet on the gut microbiota, Noriega et al. *Case Reports in Medicine*, 2016. (https://www.hindawi.com/journals/crim/2016/3089303/)

19 Mercury levels in commercial fish and shellfish, US Food and Drug Admin. http://www.fda.gov/Food/FoodborneIllnessContaminants/Metals/ucm115644.htm

20 Association between omega-3 fatty acid supplementation and risk of major cardiovascular disease events: a systematic review and meta-analysis. Rizos et al, *JAMA*, 2012

21 Meta-analysis of prospective cohort studies evaluating the association of saturated fat with cardiovascular disease. Siri-Tarino et al, *Amer J of Clin Nutrition*, 2010

22 Association of dietary, circulating and supplementary fatty acids. Chowdhury et al, *Ann Internal Medicine*, 2014

23 Meat consumption and mortality. Rohrmann et al, *BMC Medicine*, 2013

24 Egg consumption and risk of coronary heart disease and stroke: dose-response meta-analysis of prospective cohort studies. *Chen et al, BMJ*, 2013

25 Alcohol consumption and risk of heart failure: the Atherosclerosis Risk in Communities Study. Goncalves et al, *Eur Hear J*, 2015

26 Influence of red wine polyphenols and ethanol on the gut microbiota ecology and biochemical biomarkers. Queipo Ortuño et al, *Amer J Clin Nutr*, 2012

27 Identification of the 100 richest dietary sources of polyphenols: an application of the Phenol-Explorer database. Perez-Jimenez et al, *Eur J of Clin Nut*, 2010 (http://www.nature.com/ejcn/journal/v64/n3s/full/ejcn2010221a.html)

28 Role of curcumin in systemic and oral health: An overview. Nagpal et al, *Journal of Nat Science, Biol and Med*, 2013

29 Odorous urine following asparagus ingestion in man. Mitchell et al, *Experientia*, 1987 (https://www.ncbi.nlm.nih.gov/pubmed /3569485)

30 You're in for a treat: asparagus. Sugarman et al, *N C Med J*, 1985

31 Metabolomics investigation to shed light on cheese as a possible piece in the

French paradox puzzle. Zheng et al, *Journal of Agricultural and Food Chemistry*, 2015

32 Relation between consumption of sugar-sweetened drinks and childhood obesity: a prospective, observational analysis. Ludwig et al, *Lancet*, 2001

33 Sucralose promotes good intake through NPY and a neuronal fasting response. Wang et al, *Cell Metabolism*, 2016

34 Artificial sweeteners produce the counterintuitive effect of inducing metabolic derangements. Swithers et al, *Trends Endocrinol Metab*, 2013

35 Artificial sweeteners induce glucose intolerance by altering the gut microbiota. Suez et al, *Nature*, 2014

36 'The conversation: your gut bacteria don't like junk food – even if you do' (https://theconversation.com/your-gut-bacteria-dont-like-junk-food-even-if-you-do-41564)

37 Dietary emulsifiers impact the mouse gut microbiota promoting colitis and metabolic syndrome. Chassaing et al, *Nature*, 2015

38 Increased gut microbiota diversity and abundance of faecal bacterium prausnitzii and Akkermansia after fasting: a pilot study. Remely et al, *Wien Klin Wochenschr*, 2015

39 What does a three-day dietary cleanse do to your gut microbiome? (http://americangut.org/what-does-a-three-day-dietary-cleanse-do-to-your-gut-microbiome/)

40 Extremely short-duration high-intensity training substantially improves the physical function and self-reported health status of elderly adults. Adamson et al, *J Am Geriatr Soc*, 2014 (https://www.researchgate.net/publication/263859177)

41 Short sleep duration, glucose dysregulation and hormonal regulation of appetite in men and women. St-Onge et al, *Sleep*, 2012

42 Fiber and saturated fat are associated with sleep arousals and slow wave sleep. St-Onge et al, *J Clin Sleep Med*, 2016

43 Neural correlates of mindfulness meditation-related anxiety relief. Zeidan F et al, *Soc Cogn Effect Neurosc*, 2014

44 Daily consumption of the collagen supplement Pure Gold Collagen® reduces visible signs of aging. Borumand et al, *Clin Interv Ageing*, 2014

Acknowledgements

I would like to say a big thank you to my publishers Aurea Carpenter and Rebecca Nicholson, and the rest of the team at Short Books for their support and wise editorial advice. Also thanks to Caroline Barton for her excellent culinary contributions and to my agents Sophie Laurimore and Andrew Nurnberg for helping me get through some challenging situations. Above all, I would like to thank Clare, who has done so much to make this book possible.

Michael Mosley trained to be a doctor at the Royal Free Hospital in London. After qualifying, he joined the BBC, where he has become a well-known television presenter. He is the author of the internationally bestselling *Fast Diet* and *The 8-week Blood Sugar Diet.* He is married with four children.

Dr Clare Bailey, wife of Michael Mosley, is a GP who has pioneered a dietary approach to health and reducing blood sugars and diabetes at her surgery in Buckinghamshire. She is the author of *The 8-Week Blood Sugar Diet Recipe Book*, published in 2016.

Tanya Borowski is a nutritional therapist who focuses on Functional Medicine. She has a diploma in Nutritional Therapy, full certification through the Institute of Functional Medicine and specialises in treating digestive complaints such as IBS, gluten sensitivity and coeliac disease.